Soviet Forest Industries

Westview Replica Editions

The concept of Westview Replica Editions is a response to the continuing crisis in academic and informational publishing. Library budgets for books have been severely curtailed. Ever larger portions of general library budgets are being diverted from the purchase of books and used for data banks, computers, micromedia, and other methods of information retrieval. Interlibrary loan structures further reduce the edition sizes required to satisfy the needs of the scholarly community. Economic pressures on the university presses and the few private scholarly publishing companies have severely limited the capacity of the industry to properly serve the academic and research communities. As a result, many manuscripts dealing with important subjects, often representing the highest level of scholarship, are no longer economically viable publishing projects--or, if accepted for publication, are typically subject to lead times ranging from one to three years.

Westview Replica Editions are our practical solution to the problem. We accept a manuscript in camera-ready form, typed according to our specifications, and move it immediately into the production process. As always, the selection criteria include the importance of the subject, the work's contribution to scholarship, and its insight, originality of thought, and excellence of exposition. The responsibility for editing and proofreading lies with the author or sponsoring institution. We prepare chapter headings and display pages, file for copyright, and obtain Library of Congress Cataloging in Publication Data. A detailed manual contains simple instructions for preparing the final typescript, and our editorial staff is always available to answer questions.

The end result is a book printed on acid-free paper and bound in sturdy library-quality soft covers. We manufacture these books ourselves using equipment that does not require a lengthy make-ready process and that allows us to publish first editions of 300 to 600 copies and to reprint even smaller quantities as needed. Thus, we can produce Replica Editions quickly and can keep even very specialized books in print as long as there is a demand for them.

About the Book and Author

Soviet Forest Industries

Peter Blandon

Forest industries represent an important segment of the Soviet economy, accounting for five percent of the total industrial output and providing essential materials for other key industries. This book, the first in-depth study of Soviet forestry in the English language, looks at the organizational, planning, economic, and technological aspects of the industry. Mr. Blandon first discusses labor force trends and geographical features of the country's forest resources, then assesses the forest industry's modernization program. He analyzes the past performance of capital investment in the industry and makes projections about the effects of future investments in order to estimate the Soviet Union's future timber output. Throughout the book considerable attention is devoted to the Soviet planning system and its influence on decision making.

Peter Blandon is a senior lecturer in economics at the City of London Polytechnic.

Soviet Forest Industries

Peter Blandon

Routledge
Taylor & Francis Group
LONDON AND NEW YORK

First published 1983 by Westview Press

Published 2019 by Routledge
52 Vanderbilt Avenue, New York, NY 10017
2 Park Square, Milton Park, Abingdon, Oxon OX14 4RN

Routledge is an imprint of the Taylor & Francis Group, an informa business

Library of Congress Cataloging in Publication Data
Blandon, Peter, 1949-
Soviet forest industries.
(A Westview replica edition)
Bibliography: p.
Includes index.
1. Forest products--Soviet Union. I. Title.
HD9765.S652B52 1983 338.1'7498'0947 83-10306

ISBN 13: 978-0-367-28820-4(hbk)

Contents

Tables

Figures

Acknowledgements

As an undergraduate I had to study a foreign language as part of my studies in forestry. I chose Russian and it was this choice that sowed the interest in combining a study of forestry with that of Russian.

During my stays in the Soviet Union there were many people who helped and encouraged me in what were, occasionally, very difficult times. Mr. Laskarev, who taught me Russian while I was at the Kirov Forest Academy, was always ready to help me, and I must especially thank Sasha Plastinin who often went out of his way to explain points of vocabulary or the workings of the forest industries during what was a very busy time for him. His help and encouragement maintained my interest in the subject.

Of the specialists I met and spoke to, special thanks must go to Professor Lobovikov and Dr. Petrov, both of whom helped to organise my studies and who spent a great deal of their time explaining fundamental points to me. Also I should thank Mr. Kochegarov, Mr. Murashkin, Mr. Ovchinikov and Mr. Rodigin who spoke to me about their areas of specialization.

It should be pointed out, however, that the projections and opinions contained and explained in this book were formulated, in the main, after my last trip to the Soviet Union. Most, if not all, of the people I met in Russia would take a more optimistic stance than the one I have taken and many would no doubt disagree with points of detail.

Back in England, I must thank Elizabeth Blandon for her invaluable help in checking the manuscript, even if it did prove that her spelling ability is better than mine. I should also apologise to my parents who worried greatly during my trips behind the Iron curtain.

Also I should like to thank Andrew Sumner and Mariko Yamawaki for their time in preparing the illustrations.

Although I am in many people's debt it should go without saying that the responsibility for any mistakes in the figures, opinions, and projections in this book is mine.

Peter Blandon

1
Introduction

More than 3.5 million trees are felled during the average working day in the Soviet Union. This amounts to about 350 million cubic metres of timber a year, or about 15 per cent of the global annual timber harvest. The extent of the forests in the Soviet Union is also impressive when viewed from a global position. About one-third of the world's timber is situated within the borders of the Soviet Union and over half of the coniferous forests of the world grow there.

It is not surprising, therefore, to find that the forest-based industries are an important branch of the Soviet economy. In 1970 about 2.8 million people were employed in the logging, woodworking and pulp and paper industries and an additional 433,000 were engaged in sylvicultural operations. This represents about 3.5 per cent of the total state employed work-force. Forest products account for about 5 per cent of the industrial output of the Soviet Union and constitute an important input into other areas of the economy. Most notable of the recipients of the output of the forest products industries are the various branches of the construction industry. Sawn timber and panel products are important as construction materials in their own right and also as shuttering, etc. in the construction of concrete panels and pillars in more modern approaches to building. Also, timber is an important input into the mining industry and in the building of railways. As well as these more direct uses of timber, the pulp and paper industry needs a large timber input. Overall it has been estimated that, if the present technology of conversion remained unchanged until the year 2000, over 1000 million cubic metres of wood would be required as an input into all of the other branches of industry (Varaksin and Stupnev, 1974).

So the forest-based industries represent an important link in the Soviet economy and so could also represent a constraint on the growth of that country's industrial output. Whilst the technology of the industries dependent upon timber changes - more concrete and less timber is used for construction, concrete sleepers are substituted for wooden ones in railways, coal and oil are used instead of firewood - and although the timber conversion industries themselves are becoming more efficient - it is probably still true to say that the amount of timber that can be felled by the logging industry represents an important internal constraint on the future development of the Soviet economy.

As well as the internal implications of the performance of the forest industries there is a significance for the other countries of the world. The Soviet Union is one of the four major exporters of round and sawn coniferous timber and accounts for about 15 per cent of the world market in these products. In this respect, the vastness of the Soviet Union's timber resources, coupled with the size of the world's tropical forests, led some commentators some years ago to conclude that there was a relatively rosy future for forestry from a global point of view. The world's timber needs could be supplied by exports from the Siberian and Amazon areas. In recent years this view has changed and now there is a more general realisation that a more active forest management policy on a world-wide scale is required - see for example the World Bank Sector Policy Paper (1978) and Eckholm (1979).

The purpose of this book is, therefore, two-fold. First, it presents a detailed study of the performance of the Soviet forest industries over the last few years and, based on that analysis, presents a prognosis of output from Soviet forests. In this respect many of the problems now seriously affecting the whole of the Soviet economy are apparent within the forest industries. There are acute problems of labour supply and, in response to this, investment plans are now in full swing in an attempt to increase labour productivity. These plans are of interest to foresters in their own right but, as they are being influenced and distorted by the nature of the planning system in the Soviet Union, their study is of general interest to all those with an interest in the workings of the Soviet economy. It is hoped that this book will accomplish the task of describing the problems faced by this sector of the economy and of showing how the problems themselves and the effectiveness of the proposed solutions are related to the planning system. This is the most important aim of the book, but there is an underlying, secondary purpose; that is to show, as a result of the prognosis of output from the Soviet Union's forests, that the apparent wealth of timber resources in that country is just that - an *apparent* wealth. The general conclusion will be that, despite the vast areas of forests alluded to above, output of timber is unlikely to increase greatly and may actually decrease over the period to the year 2000.

A Background and a Brief Resumé of Results

Some general figures describing the performance of Soviet logging and the forest products industries in the past are shown in Table 1.1. A brief look at the figures describing logging output shows that in the recent past the performance of this sector has been far from impressive. Although output increased by 333 million cubic metres between 1913 and 1975, an increase of more than five times, in recent years output has increased only slowly and between 1955 and 1975 the volume logged in the Soviet Union grew by less than 1 per cent per annum. However, this figure underestimates the true potential of the forest products industries for expansion for during that period there was a decrease in the proportion of firewood in the total harvested. The percentage of industrial wood, i.e. all wood except firewood, in the total volume harvested increased from 63.5 per cent to 82.2 per cent in the twenty years from 1955 to 1975

TABLE 1.1 Performance of the Forest-based Industries

Year	Logging output*	Manufacture of sawn timber*	Manufacture of plywood**	Manufacture of particle board**	Manufacture of fibre-board***
1913	62.0	14.2	203.5	-	-
1922	57.8	3.1	10.0	-	-
1928	61.7	13.2	185.2	-	-
1932	164.7	24.4	423.6	-	-
1940	246.1	34.8	731.9	-	-
1945	168.4	14.7	192.2	-	-
1950	266.0	49.5	657.5	-	-
1955	333.9	75.6	1049.0	34.0	15.2
1960	369.0	105.6	1353.5	161.0	67.7
1965	378.9	111.0	1756.1	800.0	138.0
1970	385.0	116.4	2045.1	1994.0	208.3
1975	395.0	116.2	2196.0	3994.0	408.0

* figures in millions of cubic metres
** figures in thousands of cubic metres
*** figures in millions of square metres

Sources: Anon, 1977a,b,c; Glotov, 1977

(Anon, 1977a). This means that the output of industrial wood increased from 212 to 325 million cubic metres - an annual average increase of just over 2 per cent.

The average growth rates, worked out over a twenty-year period, hide an important fact. That is that the rate of growth of output from the Soviet Union's forests has been falling. In the decade 1955 to 1965 the volume of timber felled grew annually by about 1.3 per cent whilst in the next ten years to 1975 the rate of growth was less than one half of 1 per cent. The main reason for this has been an unimpressive growth in labour productivity and a decline in the number of workers willing to work in the forestry profession. During the twenty years from 1955 the productivity of logging workers increased by a little over 4 per cent per annum and the actual number of people employed fell by about the same amount; so much of the investment in new equipment during that period and the increases in productivity that it made possible were absorbed, so to speak, by a decline in the number of workers in forestry.

This effect continues to the present day. It is still very difficult to attract workers to logging and so the major objective of investment in this branch of the economy has been to increase the productivity of labour - even if the increased capital required to effect this is very expensive and causes costs to rise. But it seems that the capital inputs are not being as effective as they might be in raising productivity and the plan for 1981 looked forward to an output from the logging industry of only 342.7 million cubic metres - a reduction from the record 1975 figure of 13 per cent (Kirin, 1981). As well as this the targets for labour productivity growth in the eleventh five-year plan of between 16 and 18 per cent over the five-year period, or about 3.2 per cent per annum, represent a reduction compared to the figures for the previous five-year plan. The tenth five-year plan from 1976 to 1980 had demanded an increase in labour productivity of 25 per cent, or 4.6 per cent per annum. This target proved to be optimistic - in 1979 productivity rose by only 1.7 per cent (Nemtsov, 1981) and, according to at least one expert, the productivity of labour in logging actually fell during the five-year period (Solomonov, 1981). Recently it has been stated that "... in spite of the growth of technology ... many enterprises are not fulfilling their plans ... In the years of the tenth five-year plan (1976-1980) 53 million cubic metres of timber were not harvested ..." (Alekseyev, 1982).

In Chapters 4 and 5 the questions of the development and introduction of new technology will be discussed. In Chapter 4 the history of logging technology will be examined in an attempt to discover features which will help in an assessment of the potential of present and future investment plans. These plans will be the subject of Chapter 5, the conclusion of which will be a forecast of possible productivity increases in Soviet logging. Briefly, this forecast is that the productivity of labour is unlikely to increase by more than 2 per cent per annum in the future. In Chapter 6 the labour side of the problem will be examined and possible future changes in the size of the labour force in forestry will be suggested. Together with the forecasts of Chapter 5, this leads to a prognosis of the volume of total removals from the country's forests. It appears that, if the analysis presented in Chapters 4, 5 and 6 is correct, logging output from the Soviet Union's forests is unlikely to increase by any more than about 1 per cent per annum.

The environment that Soviet forestry finds itself in is also an important factor in the consideration of the possible future for the Soviet forest industries. In this context *environment* has two meanings - one, the nature of the forests, timber and locale in which they occur and two, the economic system in the Soviet Union. In its first meaning the environment will be examined in Chapters 2 and 3. In Chapter 2 the nature of the Soviet Union's forests will be discussed and a general geographical background to the forest industries will be presented. These aspects of the situation facing Soviet forestry are not really very favourable. The quality of the growing stock to be logged is declining, average tree sizes and stocking rates are falling. The harshness of the climate affects not only growth rates but is also having organisational effects on logging concerns and is reducing the ability of the sector to attract labour.

In its second meaning, a consideration of the environment that the Soviet forest industries find themselves in requires the economic system to be considered. This will be mentioned throughout the text when appropriate but Chapters 7 and 8 will specifically address themselves to this point. In Chapter 7 the planning system and the related price system will be examined. The planning system affects forestry and logging in two basic ways; first directly in the sense that the forest industries receive plans that they have to fulfil and, secondly, the inputs they receive are produced by industries which themselves have plans to fulfil. In its first aspect, the planning and price system conspire to build inefficiency into logging operations. The price of hardwoods is so low that in many cases, operating to maximise the value of the output of a logging concern, managers find it rational to leave hardwood species standing after logging over a site and move on to fell coniferous species on nearby sites. From the second point of view, the input into forestry that is of major importance, capital equipment, is itself the output of an industry that is planned in a similar way to the forest industries. This aspect will be discussed in Chapters 7 and 8 as well as in passing in Chapter 5. The effects of the planning system on the capital goods industries are to create incentives to produce 'old' rather than 'new' types of equipment and, secondly, to create disincentives for the production of spare and replacement parts. These two effects mean that the investment programmes in logging have had a disappointing effect on overall productivity levels. In Chapter 8 two specific economic problems will be examined in detail. First, investment appraisal methodology will be reviewed and it will be seen that the methodology is not completely satisfactory in indicating the type of logging system which represents an 'optimum' capital intensity. Secondly, the definition of an optimum rotation length will be examined.

Possible Responses to the Situation

The arguments to be presented in Chapters 2 to 8 contain the main bulk of the book and provide the basis for the rather pessimistic predictions about the future output from the logging industry.

The remaining two chapters, Chapters 9 and 10, deal with the possible responses to the situation.

There are two areas in which responses to this can be made - within the logging industry itself and outside the logging industry. Responses within the logging industry will be discussed in Chapter 9. The main way in which an inability to expand output from the Soviet Union's forests can be got around is to attempt a fuller utilisation of the timber felled. It is possible to achieve this in two ways. One, an experimental one as yet, is to extract stumps for conversion. This is not expected to account for a significant volume of output in the near future despite the fact that experiments in this field began in the 1960s. The second way, however, is becoming increasingly important; this is to use timber which would previously have been left unused or used as firewood - tops of timber, branches and logging waste - to produce chips for use in the processing industries. It has already been noted that the percentage of firewood in the total volume cut has declined continuously since the Second World War. In 1945 about 60 per cent of the timber logged was not used for conversion into some form of timber product. Most of it was used for firewood and, as will be seen in Chapter 4, some was used to fuel wood-burning generators and tractors. In 1975 the volume of wood converted into some timber product was 82.2 per cent (Anon, 1977a). Clearly this is a process that tends to a natural limit of 100 per cent and so any future increase in the output of usable timber as a percentage of total volume felled is unlikely to be as impressive as it has been in the past. This aspect of the forest industries is influenced by the planning and pricing system and so will also be referred to in Chapter 7.

Two other such internal responses that can be made to the problem of timber shortages will be discussed in Chapter 9. First there is the possibility of expanding the exploitation of the so-called Group I forests. These are forests that are designated as protection forests or forests with some amenity significance. Some of these forests are located in the European part of the country and so could be used to expand the output of timber there where much of the processing industry is located and where living conditions are better than they are in the Siberian parts of the country. This latter point may help to resolve partly the labour problem. Another way of increasing the supply of timber to the domestic economy is to divert it from the international one - that is to reduce exports. This is likely to be undesirable from the point of view of the Soviet Union's balance of payments position and its need for Western capital imports and grain, but a similar effect could be achieved by substituting higher value exports such as panel products for round timber and sawn wood provided that the price differential is sufficient.

In the final chapter, Chapter 10, the forest products industries will be discussed. It can be seen from Table 1.1 that output of sawn timber has only grown by 2.2 per cent per annum between 1955 and 1975, and in the years between 1970 and 1975 output actually fell. Plywood output increased by a slightly better 3.9 per cent per annum over the same twenty years. Both of these products seem to have been constrained to a large extent by the slow-down in the growth of output from the Soviet Union's forests. Neither are products that would accommodate the use of low quality timber easily. Output of particle and fibreboards have increased dramatically, by 27 and 18 per cent per annum respectively. There is probably some capacity for expansion here as these are panel products which can utilise the low quality wood and logging waste that will have to be used in the future if output is to be expanded. The eleventh five-year plan looks forward to an increase in particle board production of 1.5 times between 1981 and 1985, or 8.5 per cent per annum, and of between 1.3 and 1.5 times for fibreboard, or 5 to 8.5 per cent per annum.

As well as these products, timber is an important input into furniture manufacture - which takes about 6 per cent of the output of the Soviet Union's forests, and paper production - which takes about 8 per cent. To a lesser extent, timber is used within the logging industry itself and for such items as telegraph poles and for direct sale to the public. Together, these absorb about 9 per cent of the timber produced.

A Word on Sources

Whilst this book stemmed from a year's study in the Soviet Union and conversations with experts in the various areas of the forest-based industries, the figures presented in the book are, almost without exception, taken from published sources - usually Soviet publications. Clearly, the conclusions reached are going to be no better than the data used to support them and so a word on the source material is advisable.

It is probably true to say now that very little covert manipulation of the figures available in the Soviet press occurs. The figures that are available are probably as accurate as any published statistics ever are. Even if they were not, it is difficult to see how any more reliable an estimate could be made of many of the figures used in this book. In any case, many of the published figures do not show the forest industries in a good light, which could be taken as evidence of accuracy.

However, this does not mean to say that the figures in the Soviet press can be used without some words of warning. There is a disconcerting habit amongst Soviet authors to give no references to their articles and to describe data in only the most general terms. Thus, for example, a figure for the productivity of labour might be quoted of, say, 500 man-days to produce 1,000 cubic metres of timber without any explanation as to the area of the analysis, e.g. the whole of the Soviet Union, only the forests under the direct control of the main Ministry, or which workers are really being referred to, e.g. all workers including even the Minister, or only those working on the forest floor at basic operations etc. For any given measurement it is usually possible to find many figures purporting to be the true value of it, sometimes almost side-by-side in one book or article. This problem is also true of such apparently important indicators as 'volume logged' etc., as the basis for measurement is left undefined. Another feature of the figures quoted in the Soviet press is that they often take the form of percentages. Thus, one reads that the number of a certain type of tractor in use in forests has increased over the last year by 10 per cent. Unfortunately, however, the figure for the number of that type of tractor in use in the previous year is not available.

In part this is not a serious problem. It is not really of much significance if the volume logged in 1975 was 395 million cubic metres *exactly*, or if the productivity of labour in logging is 570 cubic metres per man-year or 560, or if the area of land in the Soviet Union classified as 'forest covered' is 768,774.1 thousand hectares. Indeed, in many of these cases, it is doubtful if the concepts themselves can be defined in totally exact terms.

Of greater importance are the changes in the figures from one year to the next, is productivity increasing or decreasing, at what rate is it changing? And for this purpose it is probably better to use a *consistent* series of figures rather than a collection of figures gleaned from different sources. Thus, as far as possible, whenever a single source in the technical press has offered a series of figures these have been used. The compatibility of the figures used in the text with other figures given in other sources has been investigated as fully as possible and any relevant differences noted in the text. However, the point should be borne in mind that less weight should be attached to individual figures than to the analysis presented in terms of changes of figures through time. All this does not mean, of course, that I have not taken as much care as possible in ensuring that all of the figures presented are accurate and representative of the true position in the Soviet Union.

2
A Geographical Background to the Forest Industries

THE SOVIET UNION

The Soviet Union is a vast country no matter what figures one considers. It covers one-sixth of the earth's land surface and ranks third in terms of population amongst the countries of the world. It contains about one-third of all standing and over one-half of all of the world's valuable coniferous forests. Spatially the Soviet Union is divided into European and Asian parts. The bulk of the land surface is located in Asia with the Ural mountains forming the traditional barrier between the two continents. The European part of the country covers about 5.6 million square kilometres from the Finnish, Polish, Czech, Hungarian and Rumanian borders in the west, Turkey and Iran in the south to the Ural mountains in the east. The Asian part covers 16.8 million square kilometres and accounts for the entire northern Asian land mass, with its southern borders shared with Iran, Afghanistan and China.

Politically the Soviet Union is a federation divided internally into a number of Republics. By far the largest of these is the Russian Soviet Federation of Socialist Republics, the RSFSR, which covers 76.3 per cent of the land area of the Soviet Union. It is also the most important Republic in terms of timber resources, containing over 94 per cent of the forest area of the Soviet Union's state forests. The population in the late 1970s of the RSFSR was 134.7 million, or 52.7 per cent of the country's total. The Republic is itself further subdivided into a number of economic regions. These are the Central region, the Volga-Vyatka region, the Central Black-earth region, the Povolzhski region, the Northern Caucasus, the Urals and the North-west region - all of which may be considered to be in the European part of the RSFSR - and Western Siberia, Eastern Siberia and the Far East - in Asia. The other Republics which make up the rest of the Soviet Union are the Ukraine, White Russia, the Baltic States of Lithuania, Latvia and Estonia, Moldavia, Armenia, Azerbaijan and Georgia - in the European part of the Soviet Union - and Uzbekistan, Kazakhstan, Kirghizia, Tadzhikistan and Turkmenistan in Asia.

All of these are shown along with the major urban centres in Fig. 2.1.

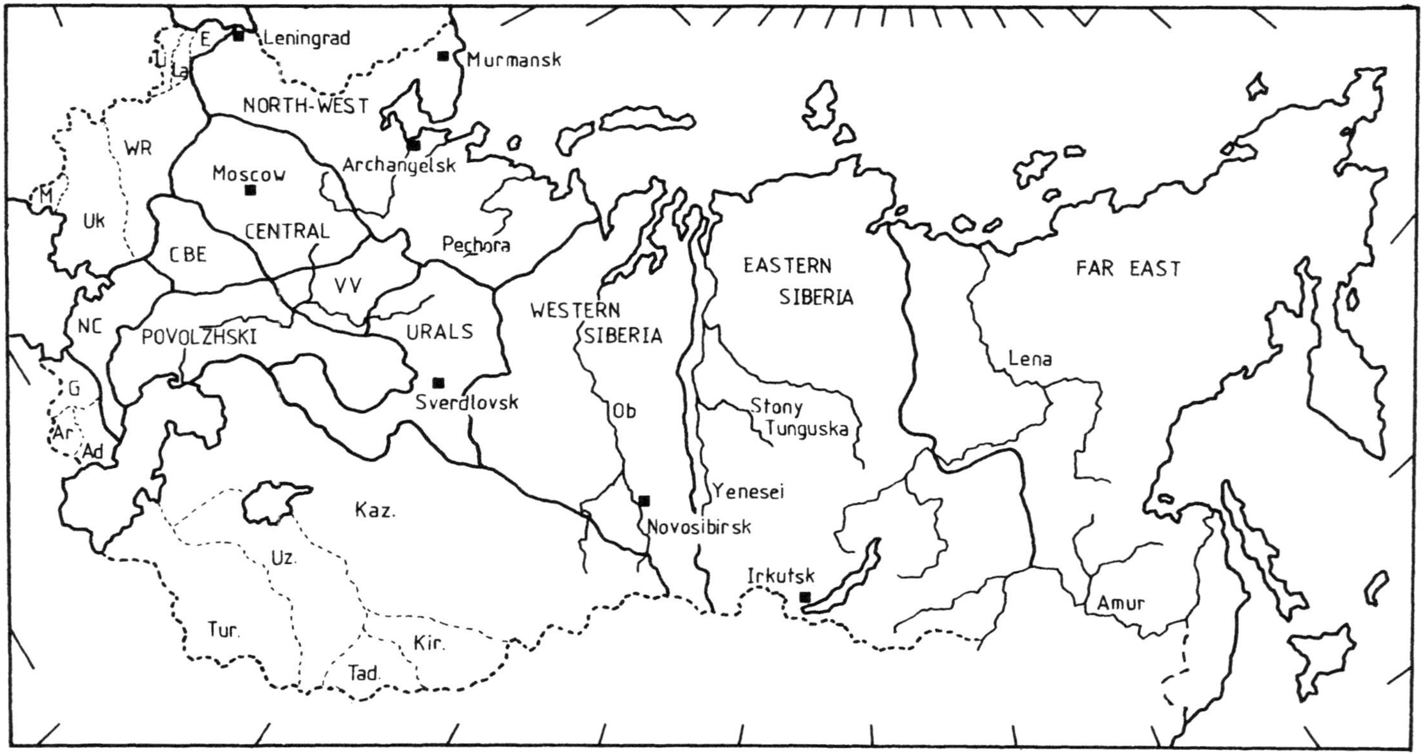

FIG. 2.1 The Soviet Union

Climatic Conditions of the Soviet Union

Temperature. The Soviet Union is famed for the harshness of its climate. This reputation is justified as the remoteness from moderating maritime influences means that in only a few places are average January temperatures above zero Centigrade - these places being around the Black Sea and in Soviet Central Asia. The north Atlantic's influence is only really felt in the very north-west of the country. In parts of north-eastern Siberia the mean January temperature can be below -40°C and in some areas the average temperature only reaches zero in mid-May and falls below it as early as mid-September (Patyakin, 1973).

The cold temperatures recorded over much of Siberia mean that all of the northern parts of the Soviet Union east of the Urals undergo as many as 200 days of snow cover each year. This, in the context of the relatively open nature of Soviet forests with their low stocking, leads to a reduction in productivity as snow has to be cleared from around the trees to prevent leaving high stumps when logged. This is counteracted to a certain extent by the fact that in the periods of hard frost the extraction of timber on ice and snow roads is possible.

The winter temperatures also mean that many of the Soviet Union's rivers freeze for considerable periods of time. For example the Ob-Irtysh river system is open for less than half of the year at its mouth. Krasnoyarsk, on the Angara-Yenisei river system, is also open for only about half of the year. Together these river systems transport more than 90 per cent of the timber felled within their basins (Lebedev and Pominova, 1974 and Patyakin, 1973). The spring thaw often brings floods to these rivers as the head waters sometimes thaw before the lower reaches.

In July the isotherms run roughly east-west and even the very north of the country experiences temperatures of 12°C. This wide range of temperatures leads to problems that have an influence on forestry. The severity of winter means that, during the summer months, the sun's warmth succeeds only in thawing the surface layers of the soil leaving the lower layers frozen. The moisture from the surface lies upon the frozen subsoil causing marshy conditions, or even flooding. This occurs in varying degrees of intensity in almost all of the eastern parts of the Soviet Union. It has been estimated that only about 7 per cent of the country's forests are left unaffected by the thaw and are accessible to the machines harvesting and transporting timber throughout the year (Vingorov, 1972). This imposes an uneven pattern on logging operations.

Also, the pattern of freezing and thawing causes the breakdown of the soil particles in the upper horizons of the soil. In some areas wind causes erosion leaving only degraded soil profiles. This leads to a demand for shelter belt forestry.

Precipitation. The Soviet Union is a relatively dry place. The annual precipitation in most parts of Siberian Russia is less than 60 centimetres and in European Russia it is only slightly higher. Much of this falls as snow. Usually rainfall in the European part of Russia results from the westerly wind and so declines the further east one goes. European Russia is also an area of fairly seasonal rainfall with most rain falling during the summer months. In the Asian parts of the country the rainfall is influenced by the monsoon cycle in southern and south-east Asia. The prevailing winds in the summer become south-easterly and, with these on-shore winds, the Pacific coast undergoes a 'rainy season' which diminishes in intensity the further one goes inland.

Parts of the Soviet Union are desert, for example, the areas in Soviet Central Asia around the Aral Sea. But more importantly for forestry, areas of Russia not classified as deserts undergo fairly frequent 'drought' years. Areas in the southern European parts of the RSFSR experience such dry periods fairly often and these are frequently associated with winds blowing into the area from the dry regions west of the Caspian Sea.

Winds. Two types of wind are of interest to forestry. The first, the *sukhovi*, has been briefly mentioned above. When these winds blow over the Steppe area of southern Russia the humidity of the air falls rapidly. When this is coupled with low rainfall the soil often becomes dry enough to allow the topsoil to be blown away. Vegetation is a means to prevent this and so there is a demand for forest belts in the Steppes to reduce this wind erosion of the soil. Pictures in Soviet books on the subject, for example Konstantinov and Struzer (1974), show young trees almost covered with wind-blown topsoil. Also, of course, the desiccating wind is injurious to all plant life and so has serious agricultural consequences; thus, the provision of shelter belts in these areas has an agricultural significance.

Part of the reason for the detrimental effect of these winds is thought to be the deforestation of European Russia in the past. This has led to a lowering of the water level in many of the rivers in the area in the summer. Between 1929 and 1956 the water level in the rivers of the south-western parts of Russia fell by about 2.5 metres (Algvere, 1966).

Another aspect of the winds of interest to forestry is to be seen in the very north of the country. The prevailing winds in the spring and summer months, and sometimes during the winter, blow from the Arctic areas southwards across Siberia. To reduce the effect of this on the climate of the northern part of the Soviet Union the forests in the most northerly parts of the country are excluded from logging operations. Belts of up to 150 kilometres in width are formed.

VEGETATION ZONES OF THE SOVIET UNION

Figure 2.2 shows a division of the Soviet Union into various zones of natural vegetation. This classification comes from Tseplayev's book on Soviet forests (1965) which has been translated into English. In Fig. 2.2 the Soviet Union is divided into six zones - each of which are, of course, capable of being further subdivided. These zones are the tundra, the forest zone, the Steppe, the desert zone, the subtropical zone and the mountainous areas. These run in rough east-west bands with the order in the list above describing the progression from north to south.

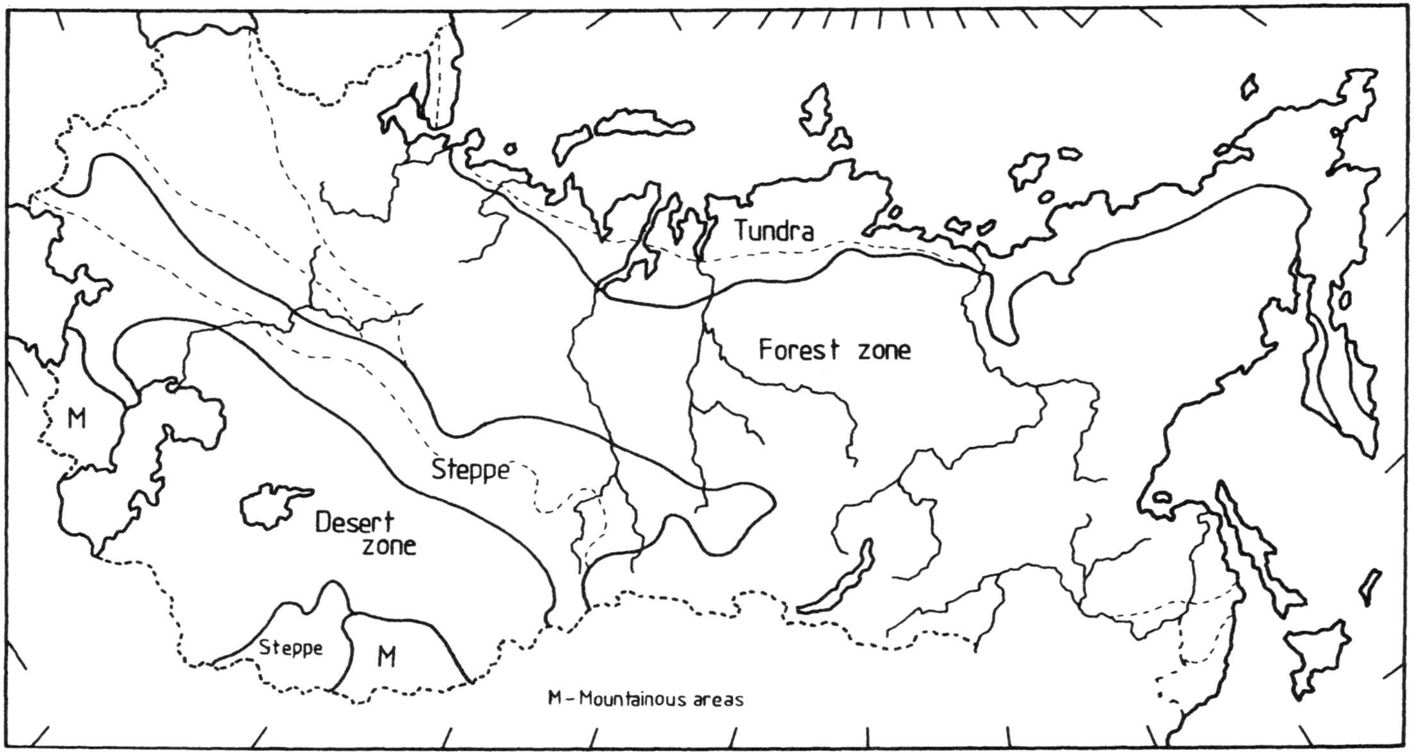

FIG. 2.2 Vegetation Zones of the Soviet Union

The Tundra

The tundra stretches from the north of the Soviet Union until it merges with the forest zone in the south; it is a vast treeless area with a plain-like topography. It has a long cold winter and a cool short summer. In the southern part of the tundra there is a sparse cover of grasses, lichen and mosses and small shrubs but the growing season is very short, from two to three months. In this period the upper soil layers thaw out and often become very boggy because of the poor drainage which characterises much of the tundra.

The transitional subzone between the tundra and the forest zone is known as the forest tundra. The width of this belt varies from 20 to 200 kilometres and it penetrates the tundra proper along sheltered valleys and in other areas where local features act to moderate the climate. The shallowness of the soil layer that thaws out in the summer period means that the root systems of the trees that grow in this area are spread out to a distance of two or three times the height of the tree. This gives the trees the necessary stability to withstand the winds that are often strong. The crown form of the trees in this subzone is often flag-like as a result of the winds that blow from the north.

The species found in the forest tundra are usually pines, spruces, larches and birch. The growing stock of these forests is very low, between 10 and 20 cubic metres per hectare, although some spruce stands have reserves of up to 60 cubic metres per hectare. Overall, the timber in this area is of little direct value to Soviet forestry and most of the forests are left as protective 'wind breaks' to reduce the effects of northerly Arctic winds on the interior of the country.

The Forest Zone

The forest zone stretches from the forest tundra in the north to the Steppe in the western parts of the Soviet Union or the actual southern border of the country in the east. The area of the zone is very large, being more than 2,300 kilometres from north to south at its widest point around Lake Baikal. In general the zone is characterised by poor soils of a podzolic nature and by areas of abundant surface water where precipitation exceeds transpiration and evaporation. The forest zone may be divided into three subzones. From north to south these are the coniferous forest or taiga, a mixed forest subzone and a belt of broadleaved forests.

The taiga. The most northerly part of the forest zone, the taiga, is mostly covered by coniferous forests with some birch and aspen. It covers an area of more than 9 million square kilometres. The subzone is characterised by low winter temperatures and relatively warm summers with temperatures above $10^{0}C$. But because of the northern location of this area and the relative flatness of north Russia, the colder Arctic air often penetrates, causing frequent frosts in late spring and early autumn which effectively reduce the length of the growing season. This is largely the reason for the relative absence of broadleaved species in the subzone (Boldanov, 1974 and Tseplayev, 1965). Podzols predominate but become less common towards the south. The trees found in this area are spruce, pine, fir, birch and aspen, with larch in the Asian parts of the taiga. The actual location of these species will be discussed later.

The taiga shows great internal variability; this may be seen in the temperatures of different areas within the subzone. The January temperatures in the east of the European part of the taiga are, on average, between -10^{0} and $-20^{0}C$ whilst in the west they are between -4^{0} and $-12^{0}C$. The number of days when the temperature is greater than $10^{0}C$ varies from 50 to 160 depending on the part of the taiga being considered (Boldanov, 1974). The average annual rainfall lies between 450 and 650 millimetres. The Siberian taiga shows a greater continentality of climate and slightly greater rainfall in its Pacific coastal areas than that exhibited by the European part of the subzone.

Considering the area in Western Siberia around the Ob river it is possible to identify five zones. In the north, bordering the forest tundra, there is a band of spruce and larch; average growing stocks are only 35 to 40 cubic metres per hectare and the mean annual increment is as low as 0.4 cubic metres per hectare. Then there is a belt of larch and pine with growing stocks of between 55 and 130 cubic metres and a mean annual increment of 0.6 cubic metres per hectare. Moving through an area of cedar and pine, a belt of birch, pine and spruce is found with considerably higher stocking rates of 100 to 250 cubic metres per hectare and a mean annual increment of about 1.5. Finally, there are the pine woods in the south of the taiga where growing stocks can reach 400 cubic metres.

About two-thirds of the area of the taiga is actually forest covered and the subzone contains about 80 per cent of the Soviet Union's timber stocks and 90 per cent of all of its valuable coniferous resources. Looking at the subzone as a whole, pine and spruce are most common in the north and the further south one moves the more important become the broadleaved species. In the southern taiga, which would include the *Kirovlesprom*, *Permlesprom* and the north of the *Lenles* logging enterprises - see Chapter 3 for the description and location of these - birch, aspen and other broadleaved species account for 37 per cent of the forest area. This is indicative of the transition of the taiga into the next subzone, that of the mixed forests.

The mixed forest subzone. The mixed forest subzone is not a belt across the whole of the Soviet Union but rather takes the form of two wedges in the European part of the country and in the Far East. This is because the continentality of the climate in the Siberian interior causes the taiga to stretch much further south until it changes fairly abruptly into the subzone of the broadleaved forests or the forest steppe.

The mixed forest subzone is characterised by less harsh conditions than those of the taiga with somewhat warmer winters and far less frequent early autumn and late spring frosts. The growing season might last for up to 190 days. The overall rainfall is relatively low - up to 650 millimetres - but is sufficient during the growing season. The soil conditions are usually podzolic and growth is quite good with stocking rates and mean annual increments at least as high as those of the southern taiga.

In the western wedge 36 per cent of the forests are made up of broadleaved species, with birch accounting for 18 per cent of the total and oak and aspen 10 per cent. Of the coniferous species, pine predominates, accounting for 53 per cent of the total and spruces account for the remaining 11 per cent of the coniferous share of these forests. In the eastern mixed forest subzone the species composition is different; there is a greater predominance of the broadleaved species - they make up 55 per cent of the forest with birch the main species again - and of the coniferous species spruce is more common than pine.

The broadleaved forests. This is a small area of the forest zone; again it is split into two separate areas rather than forming a belt across the country. The reasons for this are the same reasons that give rise to the division of the mixed forest subzone. In Europe these forests form a small area from the Carpathians to the Urals and in the east the forests grow along the Amur river. Small areas that can be included in this subzone occur in the Crimea and Caucasus. The subzone has a favourable climate for tree growth and productivity can be high. But, in Europe at least, the subzone represents an area of high population and agricultural activity and therefore the actual forested area is small.

The Steppe Zone

South of the forest zone the Steppe zone occurs. Before the Steppe proper, however, a transitional subzone (the forest steppe) is identifiable.

The forest steppe. This area has what might be described as warm summers and cold winters; the annual precipitation is between 400 and 600 millimetres but droughts are not uncommon during the growing season. The forests form a rough belt across the country from the Carpathians in the west to the border with Mongolia roughly south-east of Novosibirsk. At its northern border the forest steppe joins the broadleaved forests of the forest zone in European Russia and in Siberia it adjoins the taiga. The major species found in these forests are pine, accounting for about a quarter of the area, oak, birch and aspen. In the more temperate parts of the forest steppe oak is the major broadleaved species - accounting for over half of the forests of the zone in the European part of the Soviet Union. In the Siberian parts of the transition zone, the harshness of the climate means that birch takes over as the leading broadleaved species.

The Steppe. The Steppe is hardly an area of importance to the logging industry. However, it does have significance for forestry in the wider sense; the area has hot summers and low summer rainfall - there are only between 280 and 460 millimetres of precipitation over the whole year. Coupled with hot, dry summers, dry winds often occur causing drought and soil erosion. As the area is of considerable agricultural significance this leads to problems and a demand for protective forestry. Hence forest shelter belts are important as an adjunct to the usual agricultural practices to ensure regularity of harvests.

The Desert Zone

The desert zone is of no real importance to forestry although it covers nearly 3 million square kilometres of the Soviet Union's land surface. Some reclamation work and protective sylviculture, for example dune stabilisation, is being carried on in areas where conditions are not too severe.

The Subtropical Zone

Although most people's idea of the Soviet Union is probably that of an icy Siberian winter, the country does have a subtropical zone within its borders - albeit a small one. It occurs on the Black Sea coast of the Caucasus area and on the southern coastal parts of the Caspian. The area overall covers about 43,000 square kilometres of which between one-third and a half is forested. Productivity in these areas is very high with growing stocks up to 500 cubic metres per hectare, but the main output is not timber but other forest products. Walnuts and edible chestnuts and fruit trees are grown there, cork oak is cultivated and viticulture practised.

The Mountain Forests

Several areas of mountains in the Soviet Union are worth brief attention. There are mountainous areas in the Caucasus, the Carpathians, the Crimea and in Kirghizia and Tadzhikistan. As the mountain forests occur in distinct areas they are probably best dealt with separately.

The Caucasus. The forests of the Caucasus mountains differ markedly depending on the aspect of the slope. Zonation usually occurs at lower altitudes on the north-facing slopes. As altitude increases the forests change from broadleaved forests of oak to mixed forests of beech, spruce and fir, then a band of coniferous forests made up of spruce and fir is encountered and then the forests end with a belt of birch. The forest belts end at an elevation of between 1,900 and 2,500 metres. Overall, broadleaved species predominate, making up about 90 per cent of the forests, with oak, beech and birch the main species.

The Carpathians. Forests grow at altitudes from about 600 metres above sea level to about 1,800 metres; the lower slopes are covered with broadleaved forests of beech with some oak, aspen and hornbeam. At about 1,200 to 1,400 metres above sea level the forest changes to coniferous species with spruce and fir the main species. The climatic conditions here are excellent, high annual average temperatures of 6° to $8^{o}C$ combine with precipitation of between 800 and 1,000 millimetres per annum. This gives rise to some of the highest stocking rates encountered in the Soviet Union of up to 800 cubic metres per hectare in beech forests and 1,200 cubic metres in the best spruce and fir forests.

The Crimea. The forests of the Crimean mountains consist almost entirely of oak on the lower slopes and pine on the upper elevations with a transition zone between.

Kighizia and Tadzhikistan. The mountains in this area of the Soviet Union lie to the north of the Himalayas and contain the highest mountains in the Soviet Union. Their foothills merge with the desert zone and, as altitude increases, a semi-desert zone is encountered, followed by a belt of scrubby vegetation between 1,200 and 2,800 metres above sea level. Above this the vegetation becomes sparser and birch becomes more common. In the Tien Shan mountain range which occurs in the northern part of Kirghizia, spruce forests are common at elevations above 1,500 metres. The forests in this area are of little importance for production of timber but are of great protective significance.

THE DISTRIBUTION OF THE MAJOR SPECIES IN THE SOVIET UNION

The most common tree in the Soviet Union is the larch, accounting for 38.5 per cent of the country's forests. It occurs mainly in Siberia. Pines account for 17 per cent and spruces and fir together for almost 14 per cent of the Soviet Union's forests. Birches and aspen are the common broadleaved species, covering 16.5 per cent of the forest area and oak and beech are also present. The geographical distribution of the major species is presented in Tables 2.1 and 2.2 in the Appendix to this chapter. In referring to these tables it should be remembered that the figures refer to forests where a particular tree predominates or characterises the area. Thus the areas of the forests given in Tables 2.1 and 2.2 do not necessarily add up to the total areas of all forests in the Soviet Union or in any specific economic region.

The Major Coniferous Species

Pine forests. Pine forests are of considerable importance to the Soviet Union; they make up about 17 per cent of all of the state forests and the growing stock is an estimated 14.3 thousand million cubic metres. Pine timber is highly valued for almost all end uses. The most important species are *Pinus sylvestris*, *P. pallsiana* (or Crimean pine) and *P. sibirica*. This latter species, a five needle pine, is often referred to as Siberian cedar in the Soviet texts - *kedr sibirski* (Shimanuk, 1974).

The fact that pine, and especially *P. sylvestris*, is capable of surviving and growing in areas of relatively poor soils, frost and low rainfall means that its areal distribution in the Soviet Union is very extensive. In fact, it can grow almost anywhere except in the northern and eastern-most parts of Siberia and in the Steppe. Most of the pine timber, however, grows in the taiga. The distribution of pine forests is shown in Fig. 2.3. The quality of these pine forests varies greatly from area to area; in general, however, it is possible to say that they improve in terms of stocking and mean annual increment and become younger in age from north to south. The reason for the first two points relates to changes in climatic and edaphic factors but the fact that pine forests are younger in

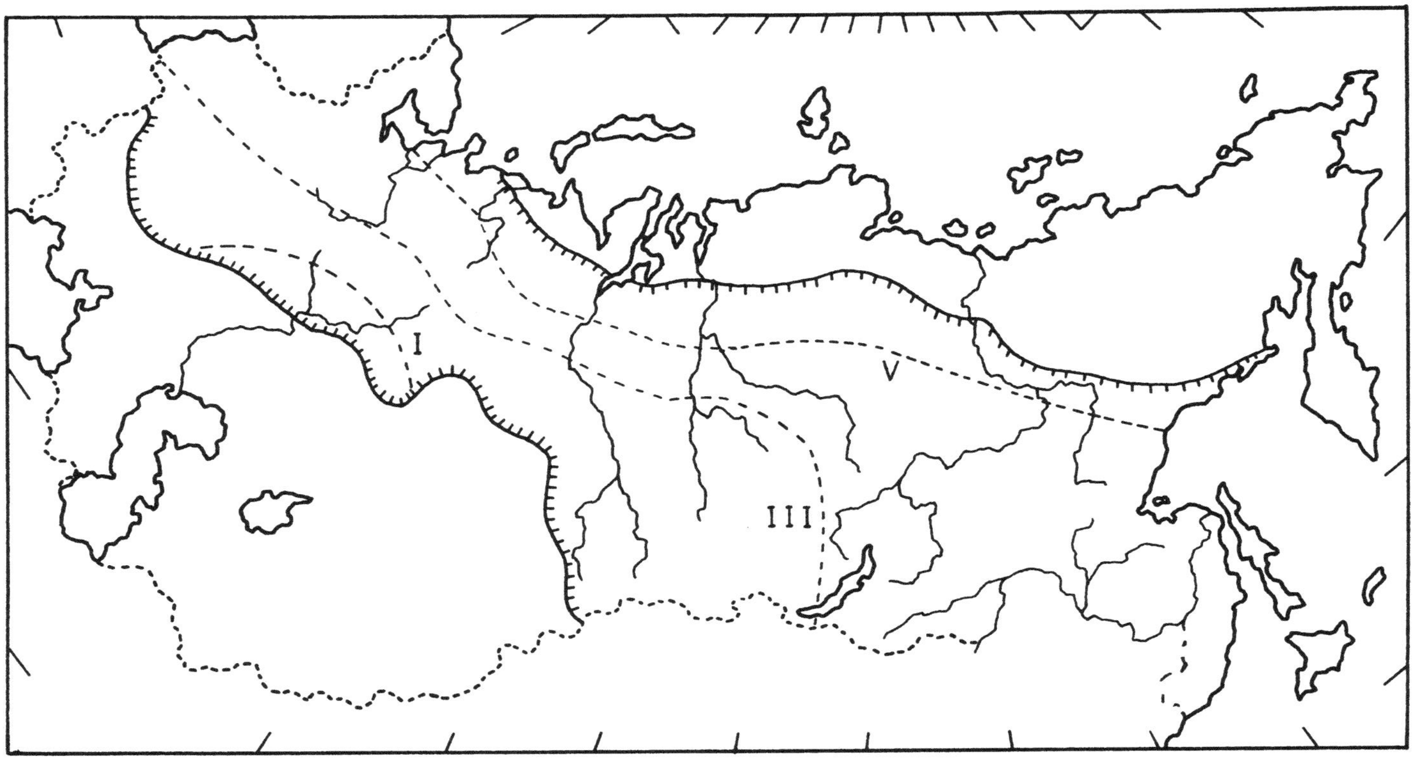

FIG. 2.3 Pine Forests of the Soviet Union

the south is more a reflection of the greater accessibility and therefore greater usage of the forests in the more southern areas of the country.

In the Soviet Union the productivity of forests is described by site classes. These are defined by the average height of the trees in the forest at certain ages. Five classes with two sub-classes are identified and these are applied separately to conifers and broadleaved trees. The classes are usually referred to by the notation Ia and I, II, III, IV, V and Va. For coniferous species at ages between 81 and 100 the average height of the trees in these various site classes would be as follows - Ia, 34 to 40 metres; I, 29 to 26; II, 25 to 23; III, 22 to 19; IV, 18 to 15; V, 16 to 13; and Va, 12 to 10 (Gatskevich, 1971). As can be seen site class Ia is the best and Va the worst. Roughly speaking, these site classes correspond to the yield classes used in British forestry - although the latter are not defined by the overall average height. For coniferous species, site class Ia is more-or-less equivalent to yield classes 14 or 16, site class II to yield classes 8 or 10 and site class IV to yield classes 4 or 6 (Hamilton and Christie, 1971). The lines indicating average site classes are also shown in Fig. 2.3 and they are aligned roughly east-west as would be expected. As they move further east they tend also to move slightly southward in line with the increased severity of the climate as it becomes more continental. The effect of the Urals is also marked, causing the 'iso-site-class' lines to be displaced southward. Thus, for any given latitude, productivity in the area of the Ural mountains will be lower.

There is a good relationship between the diameter of pine and the site classes just described. At age 90 the relationship is as follows. For site class Ia the average tree diameter at breast height would be about 37 centimetres; for site class I, 33 centimetres; II, 28; III, 24; IV, 20; V, 15 and for site class Va about 10 centimetres (Tseplayev, 1965). As the average rotation length for pine is usually less than 120 years and might be as short as 80 years, and as a site class of III represents quite high productivity in the Soviet Union pine timber in the future is unlikely to grow any more than 24 centimetres in diameter in the most important logging areas of the country.

This figure is of significance for the plans for the mechanisation of logging and for labour productivity. The figures above suggest a low average tree volume for the taiga's pine forests. With a site class of IV, say, at age 90 the average height will be 16.5 metres and the average breast height diameter 20 centimetres. This would give an average tree size of about 0.26 metres - on the assumption of a form factor of one-half. This compares with the lowest yield class for Scots pine used in British forestry - yield class 4 (Hamilton and Christie, 1971). This low average tree volume reduces the productivity of the new logging equipment now being introduced and so the shift of logging operations into the more remote and less productive parts of the country has implications for the long-run potential of Soviet forestry to increase its output.

Regeneration of pine stands is usually quite good and there is often much regeneration after forest fires - which may be an explanation for the many stands of almost pure even-aged pine that are in existence. In the northern regions, however, the interval between good seed years may be as much as twenty-five to thirty years.

Spruce forests. Like the pines, the spruces are of considerable importance to the Soviet economy. The timber is suitable for almost all end uses and is especially favoured for pulping. About 81 million hectares of spruce forests exist, often in mixture with pine, fir or larch and birch and aspen.

In the Soviet Union there are nine species of spruce that occur naturally and several have been introduced. The most important to the economy are *Picea abies*, or Norway spruce - *Yel obiknovyennaya* or *yevropeiskaya*, *P. obovata*, *P. orientalis*, *P. schrenkiana* and *P. pungens*. Norway spruce grows in a belt in the taiga and in the mixed forests of the European part of the Soviet Union. *P. obovata* or Siberian spruce is usually found in western and central Siberia as it is better suited to a more continental climate. These are the two major spruce species and their distribution is shown in Fig. 2.4. Also *P. orientalis* grows in the mountain forests of the western Caucasus and *P. schrenkiana* occurs in the Tien Shan mountains of Kirghizia. The overall volume of spruce timber in the Soviet Union's forests is about 10.6 thousand million cubic metres.

The distribution of site classes for spruce is very similar to the pattern exhibited by the pine species of the Ural mountains and the continentality of the Siberian interior having somewhat less of an effect. Spruce often also occurs as a second storey especially in areas with better soil characteristics. Natural regeneration of spruce is more shade tolerant than a lot of species, e.g. pine. This means that in logging spruce forests there is often a requirement to preserve the natural regeneration so that the second generation forests are likewise made up of the valuable spruce species rather than less valuable broadleaved species which might take over if the area was totally cleared.

Fir forests. These species form only about 2 per cent of the Soviet Union's forests but the total growing stock of firs is about 4 thousand million cubic metres. The firs are usually found in mixture with other species - often spruce. Ten species of fir grow in the Soviet Union but the most important ones are *Abies sibirica*, *A. nordmanniana*, *A. alba* and *A. holophylla*. The distribution of the main species is also shown in Fig. 2.4. *A. sibirica* occurs mainly in Western Siberia but does cross the Urals to make up a small amount of the forests of the Komi ASSR. *A. nordmanniana* grows in the western part of the Caucasus and *A. alba* grows in European Russia in the Ukraine. The greatest concentration of fir stands are located in Georgia and in the area around Krasnoyarsk. Some fir stands are highly productive and record the highest possible site classes. In Georgia a maximum productivity of 2,200 cubic metres per hectare has been recorded with individual trees reaching two metres in diameter.

Although fir species are in general shade tolerant and so a natural undergrowth of fir below the main canopy is not uncommon, clear-felling causes this under-storey to check and the area becomes overgrown by other species. Natural regeneration of the desirable fir species is best ensured by a shelterwood system with the felling

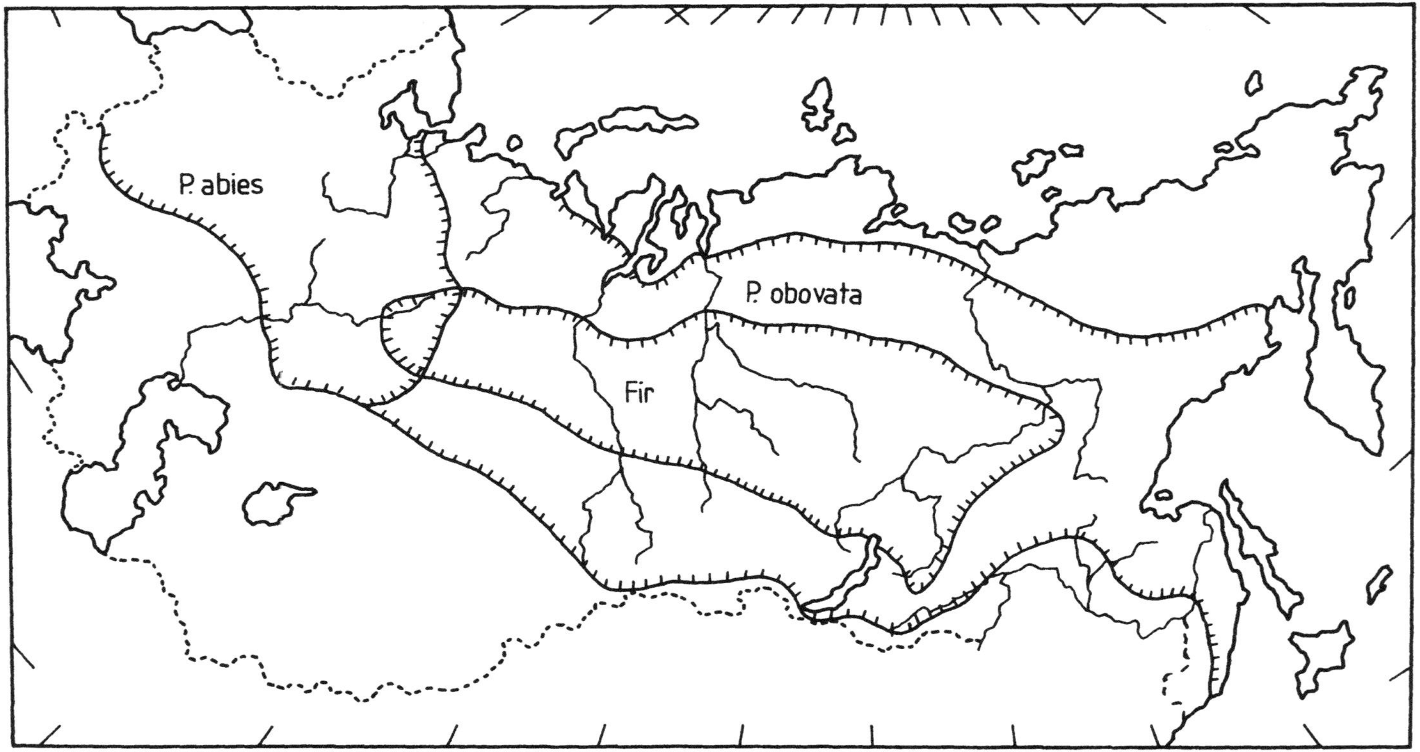

FIG. 2.4 Spruce and Fir Forests of the Soviet Union

of an area being carried out in two or three stages with intervals of between fifteen and twenty years. Failing this, any regeneration that does occur is likely to take the form of a change of species.

Larch forests. In terms of area covered, the larches are the most important group of trees in the Soviet Union and they are the main forest-forming species in Siberia and the Far East. However, it is not so important a group of species for the users of timber. There are a number of reasons for this. The timber is resinous and not popular with timber users. Also, it occurs mostly in the more remote parts of the country where the logging industry is only just beginning to develop. These areas tend to have a poor infrastructure and the timber felled in these regions has to be floated. The density of the larches, however, makes them relatively difficult to transport in this way. Thus, although overall the larches account for more than 274 million hectares of forest with a growing stock of more than 28 thousand million cubic metres, they account for only about 7 per cent of the timber felled (Polekhin and Dmitriev, 1980).

Larches are amongst the most cold resistant species in the Soviet Union and this accounts for their predominance in the Siberian regions and the Far East. Overall, naturally-occurring larch forests are not of the highest productivity; average growing stocks are between 100 and 200 cubic metres per hectare with mean annual increments of about 1.5 cubic metres. Larch has been planted in the European parts of Russia, although only on a small scale. Here productivities are much greater; growing stocks in excess of 1,000 cubic metres are possible with mean annual increments of 10 cubic metres per hectare (Tseplayev, 1965).

There are eleven naturally-occurring larch species in the Soviet Union of which *Larix sibirica, L. sukaczewii* and *L. dahurica* are the most important. Their distribution is shown in Fig. 2.5. *L. sibirica* and *L. dahurica* occur mainly in the Siberian parts of Russia and in the Far East with *L. sibirica* being limited to Western and parts of Eastern Siberia. *L. sukaczewii* stretches from just east of the Urals, overlapping slightly with *L. sibirica,* westwards across the Komi ASSR to Archangelsk and Leningrad in northern European Russia. *L. decidua* also occurs naturally but only in the western-most part of the Soviet Union in the Carpathians.

Clearly, although larches are now of only limited importance to the Soviet forest industries, their importance will have to increase in the future.

The Major Broadleaved Species

The distribution of the major broadleaved species is shown in Table 2.2 in the Appendix of this chapter.

Birch forests. Birches are very common in the Soviet Union growing everywhere except in the desert zone. About forty species of birch are naturally occurring in the Soviet Union of which the most important are *Betula verrucosa, B. pubescens* - both occurring in European Russia and Western Siberia - *B. dahurica* in Eastern Siberia and *B. costata* in the Far East. In many areas birches represent the limit of tree growth, e.g. in northern Siberia birches

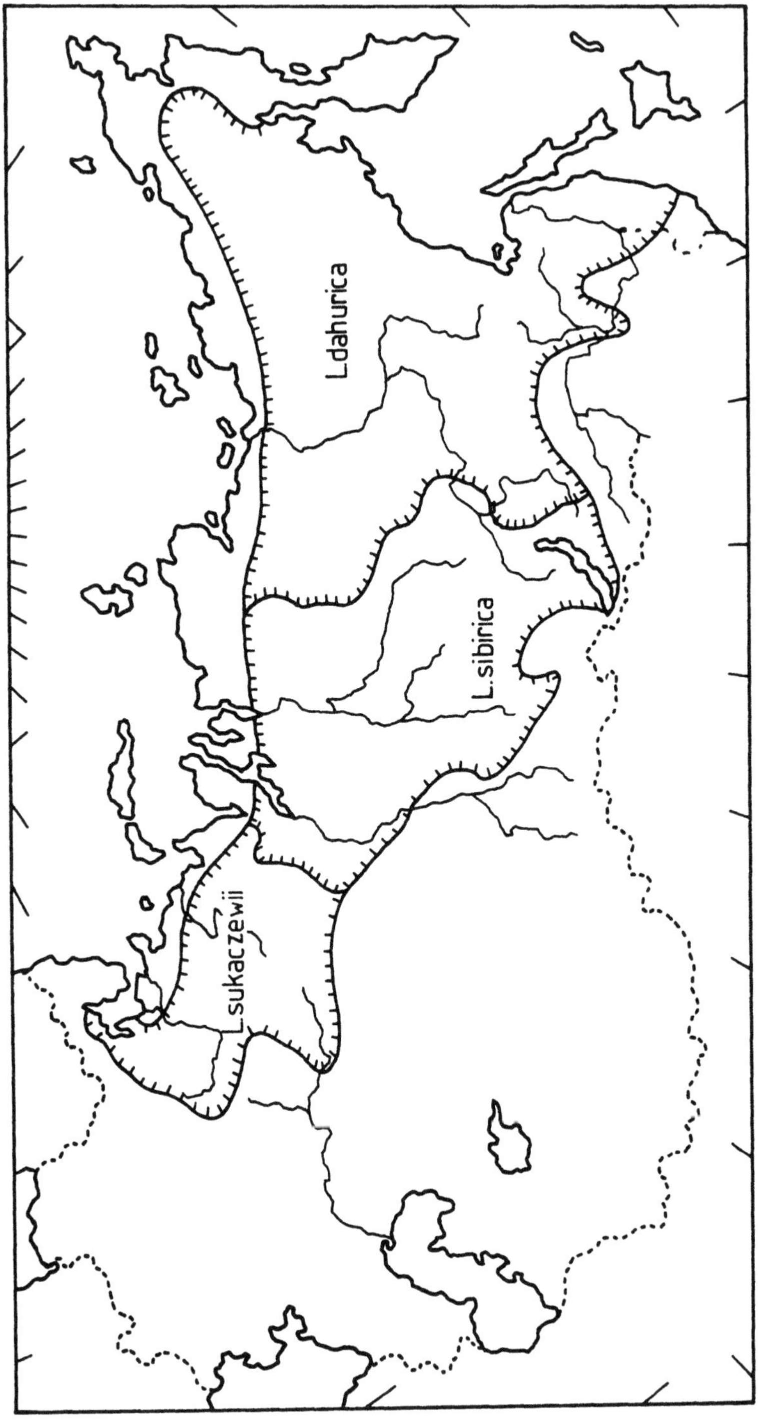

FIG. 2.5 Larch Forests of the Soviet Union

grow as far north as latitude 71°N and in mountainous regions the tree line is represented by the limit of birch growth. Overall it is estimated that there are about 7,000 million cubic metres of birch timber in the Soviet Union.

However birches are not regarded as one of the most desirable species despite their wide use in the national economy. The timber is used for plywood manufacture, for joinery and ski making as well as in the wood chemical industries. But in many areas it is still regarded as a weed species and often it is left standing after the more valuable species have been felled and extracted. This means that, very often, the logging of an area is accompanied by a change in species as the birches left are able to colonise cut-over areas very quickly. This change over has occurred to a great extent in European Russia and is now being observed in Siberian forests.

Ecologically, the birch forests are not stable and often provide shelter for more shade tolerant species such as spruce to form an under-storey; these then later become the main species for the site. However, this process is probably a fairly lengthy way to arrive at a desirable spruce forest compared with more 'artificial' alternatives of species management.

Aspen forests. Aspen is often associated with birch and it is often a mixture of birch and aspen that takes over areas of forest when the coniferous species have been removed. It is somewhat less tolerant than birch to low temperatures and so does not cover quite the same area as birch does.

Aspen, *Populus tremula*, grows quickly but is attacked by heartwood rot under most conditions when it is as young as thirty or forty years. Rotation lengths involving aspen are usually low therefore and the yield of useful timber is often as little as one third of the total volume logged. Its timber is used for match manufacture, plywood and as a raw material for paper production and the wood chemical industries. However, like birch, it is often considered to be a weed species and left standing after other timber has been removed.

Beech forests. Beech forests mostly grow in the warmer parts of the Soviet Union where the climate is also relatively humid. This is the case in the Caucasus and in the Crimea and Carpathians. Although limited in extent, beech forests are amongst the most productive in the Soviet Union with growing stocks in excess of 1,000 cubic metres per hectare and mean annual increments of up to 5 cubic metres. Three species of beech grow in the Soviet Union; these are *Fagus orientalis*, most common in the Caucasus, *F. sylvatica*, which grows in the Carpathians and *F. taurica*, a hybrid of the first two which is mostly found in the Crimea.

Oak forests. Oak occurs naturally in European Russia south of Leningrad and the Komi ASSR. Because this represents the most densely populated areas of the country the actual area of land under oak forests has diminished radically in recent historical times. The timber has been used for ship construction, furniture and vehicle manufacture, cooperage and for general construction. The use of oak for naval purposes led Peter I to pass laws to protect the oak forests; in some circumstances the law allowed the death penalty for the felling of timber suitable for ship building.

Although nineteen species of oak occur naturally in the Soviet Union the most important economically are *Quercus robur*, which grows in most parts of European Russia and in the Crimea and Caucasus, *Q. petraea*, which grows in the west of the country, *Q. pubescens*, which occurs in the Caucasus and Crimea and *Q. mongolica* which grows in the east of Siberia and the Far East if the climatic conditions are not too severe. Also *Q. suber* is grown in the Crimea and Caucasus for its bark which is used as cork. There are estimated to be about 700 million cubic metres of oak in the country, the species being most important in Moldavia where it accounts for nearly three-quarters of the forests.

FOREST RESOURCES OF THE SOVIET UNION

The statistics regarding the forest resources of the Soviet Union are quoted in different ways at different times. The basic geographical unit used varies; sometimes figures are given using a division of the country or the RSFSR into European and Asian parts. Here the Ural mountains represent the line of division. Sometimes the economic regions are used to break down the RSFSR - these were outlined earlier and are shown on the map in Fig. 2.1 - and the other Republics are used to describe the rest of the country. The two methods correspond in the following way: Asian parts of the country consist of the Soviet Socialist Republics of Uzbekistan, Kirghizia, Tadzhikistan, Turkmenistan and Kazakhstan and the Eastern and Western Siberia and Far East economic regions of the RSFSR as well as the eastern part of the Urals region. The rest is regarded as Europe. Another way of dividing the country that is often adopted is to make use of the so-called heavily forested and lightly forested areas: *mnogo-lesnoi* and *malo-lesnoi*. These correspond roughly to the northern European and Asian parts of the RSFSR on the one hand and the central and southern European parts of the RSFSR and the rest of the Soviet Union on the other. Thus, the heavily forested area includes all of the Far East and Eastern Siberia, the northern parts of Western Siberia and the Urals and all of the North-west with the exception of the areas around Leningrad and Novgorod.

Forest Areas in the Soviet Union

In a recent book by Vorobiev *et al.* (1979) the results of a survey of the forest resource dated 1 January 1973 are reported in detail. According to this survey, the total area of the land and water in the charge of the State forestry service (the State forest fund or *Goslesfond*) is 1,229,657.4 thousand hectares. Of this area, 94.5 per cent comes under the direct control of the centralised forestry system, 1.8 per cent are under the control of collective farms and the remaining 3.7 per cent are attached to other ministries. This area of *Goslesfond* is divided between Europe and Asia with 236,905.5 thousand hectares located in Europe and 992,751.9 in Asia. Of the total area about 94.5 per cent is situated in the RSFSR.

The area in *Goslesfond* is subdivided into various categories: the subdivisions are shown in Fig. 2.6. Of these, the forested and the unforested areas are the two basic ones.

FIG. 2.6 Classification of the Land of Goslesfond

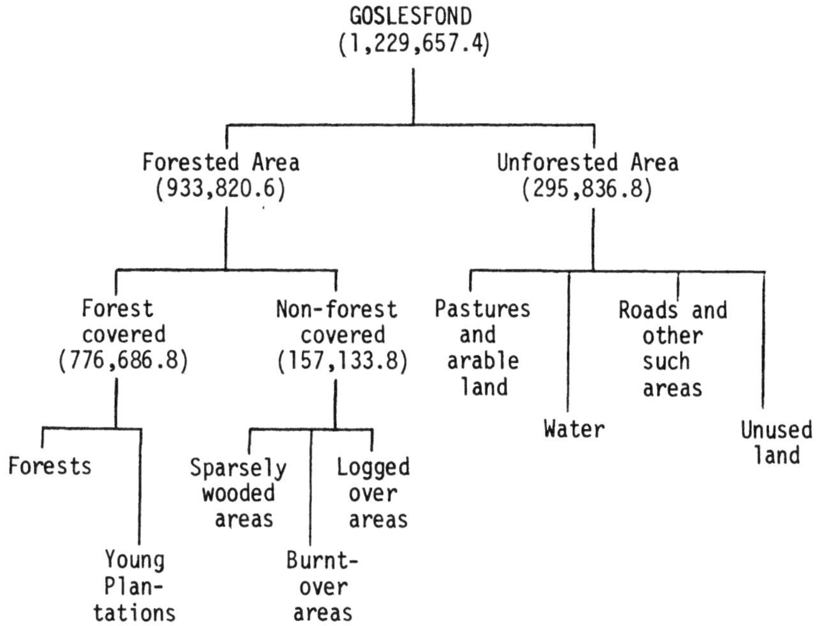

<u>The forested area.</u> About 76 per cent of the total area of *Goslesfond*, 933,820.6 thousand hectares, is classified as forested land; of this, 197,440.6 and 736,380 thousand hectares are to be found in the European and Asian parts of the country respectively. The forested area is further subdivided into forest covered and non-forest covered land. Of the total forested area of 933,820.6 thousand hectares, 776,686.8 thousand hectares are forest covered and the remaining are classified as non-forest covered. The forest covered land contains ordinary forests and areas under very young plantations. The non-forest covered area consists of very sparsely wooded land - no real definition of 'sparse' is offered in the Soviet texts - burnt-over areas, and areas that have been logged over in the recent past but have no significant growth as yet. Thus, all of the non-forest covered land has had, will have or ought to have timber on it and so is included in the overall category of forested land in the classification of *Goslesfond*.

Of the important forest covered land, 188,959.1 thousand hectares are to be found in the European part of the Soviet Union and 587,727.7 thousand hectares are in Asia.

<u>The unforested area.</u> In all, 295,836.8 thousand hectares are included in *Goslesfond* but are classified as unforested areas.

These contain pastures and arable land, water bodies (lakes, rivers, etc.) and roads and other such objects and unused land.

Details of the areas in *Goslesfond* broken down according to the system of classification are shown in Fig. 2.6 and according to their location are shown in Tables 2.3, 2.4 and 2.5 in the Appendix to this chapter.

Species Distribution

According to the classification of forests and the figures given in Vorobiev *et al.* (1979), coniferous species cover over 75.1 per cent of the forested areas in *Goslesfond*, the main broadleaved species cover 19.8 per cent and the remaining 5.1 per cent are accounted for by bushy species. The figures show a slight concentration of the coniferous species in the Asian parts of the Soviet Union, where they cover 78.4 per cent of the forested area whilst in the European part the figure is only 63 per cent.

The most common coniferous species are the larches, accounting for 38.5 per cent of the forested area in the country overall. But these are almost entirely located in the Asian part of the Soviet Union. The major coniferous species in the European part of the country are the spruces and pines. The broadleaved species of alder, aspen and birch are more common in European Russia than elsewhere, where they cover 30.5 per cent of the forested area. Of course, in the southern Republics broadleaved species account for a far greater proportion of the forests. In Georgia, Armenia, Azerbaijan, Moldavia, the Ukraine, Kazakhstan and Uzbekistan they constitute over half of the forests.

Table 2.6 in the Appendix to this chapter contains the percentages of the forested area of *Goslesfond* that are covered by the various species. It should be noted that these figures are more-or-less compatible with the figures presented by Tseplayev (1965) which are reproduced here in Tables 2.1 and 2.2. The tables would be unlikely to be completely identical as, first, they relate to different surveys taken at different times and secondly, the definition of the area covered by spruce etc. is unlikely to be the same for the authors of the two sources.

The Division of the Forests of the Soviet Union by Use

Forests in the Soviet Union are divided into three broad groups. These groups do not correspond exactly to specific Ministries or other management units, but are defined in terms of the forests' use. Thus, forests of different groups are likely to be under the control of one enterprise.

Group I forests. Group I forests account for a little less than 16 per cent of the total forest area of *Goslesfond*. They include state and agricultural forests that fulfil protective roles such as belts along rivers, lakes and reservoirs, or on slopes to prevent erosion, forests that act as shelter belts in the agricultural context, green belts and recreation forests around towns, forest parks, forests around industrial concentrations, protective belts along railways and roads to prevent snow drifting in winter and tundra forests that are supposed to prevent northerly winds from too adversely affecting the climate of the interior of the country.

The total area of such forests is 194,285.9 thousand hectares. Of this, about one-third is located in the European part of the Soviet Union and the rest in Asia. The most important categories of Group I forests, accounting for well over half of their total area, are the belts along rivers and other bodies of water and the tundra forests.

The management objectives of these forests are to create whatever service that the forest is designated for; they are managed by mainly sylvicultural considerations and such operations as clear-felling are almost totally excluded. The rotation length is longer than in most forests and the usual method of harvesting is by some form of selective cutting. Thus, although these forests cover about 16 per cent of *Goslesfond* they contributed only about 5 per cent of the total volume of timber harvested in 1975. This point will be discussed in more detail in Chapter 9.

Group II forests. These forests, making up the smallest group, are located in the main in the European part of the country. This is the lightly forested part, areas such as central European Russia and the region around the southern Urals where the forests have been over-exploited in the past. These forests are managed more intensively than most in the Soviet Union; they may be planted or sown, possibly with some prior soil preparation, and are managed throughout their life with intermediate cuttings. Thus, thinnings are becoming increasingly important as a part of the output from these forests.

The forests are managed, usually, to maximise the volume of desired timber assortments produced. They are subdivided into protective areas, similar to Group I forests but where a more productive slant towards forestry is permissible, and exploitable areas, the latter forming the bulk of the Group II forests. Some argue that, gradually, as more and more intensive sylviculture becomes common, the area of Group II forests will expand at the expense of the final group.

Group III forests. These form the bulk of the Soviet Union's forests, 77.3 per cent in all. They are the main forests used for logging and are situated in the north of European Russia, Siberia and the Far East. These forests are felled, usually by some clear-felling system with the annual cutting areas arranged spatially so that adjoining unfelled areas can seed the area logged over and so ensure natural regeneration. Again, some areas are set aside as protective belts (usually along rivers) but the main subdivisions of Group III forests are exploitable (to be used for logging) and reserve forests, which are forests which are not yet being logged, usually for reasons of their remoteness or a lack of a proper transport network.

The distribution of the forest area according to these groups is shown in Table 2.7 in the Appendix to this chapter.

The Growing Stock and Age of the Soviet Union's Forests

The growing stock of the Soviet Union's forests is in the region of 80 thousand million cubic metres; this represents more than 200 times the annual removals of timber. More importantly, the gross annual volume increment to the country's forests is estimated to be about 880 million cubic metres, 600 million of which is coniferous and 280 million of which is accounted for by broadleaved species; this is more than twice the annual removals. Breakdowns of the increment figures are given in Table 2.8 in the Appendix to the chapter.

The average increment is in actual fact relatively low when expressed on a per hectare basis and is very low in comparison to the figures achieved in plantation forestry. Part of the reason for this lies in the age structure of the Soviet Union's forests as they are predominantly mature and overmature. Only in the central European areas of the country in the Group II forests is the age structure more 'normal'. Indeed, according to most theoretical texts the 'normal' forest (one in which each age class up to maturity is equally represented) is put forward as the ideal (Anuchin, 1977a). But of the Group III forests, 85.94 per cent are classified as mature or overmature.

This might, however, be a slightly misleading figure as the definitions of mature and overmature are based on age alone and not on any technical, sylvicultural or economic concept of maturity. Coniferous species are defined as being mature if they are older than eighty years, and the most common broadleaved species are mature at forty years of age.

The details of mature and overmature timber are shown in Table 2.9 in the Appendix of this chapter; what the figures show is that the Group III forests in the Asian parts of the country have a more skewed age distribution than those in Europe. Also the Group II forests in the RSFSR do not show the same degree of overmaturity as the Group III forests do. This would be expected as these forests are those which have been over-exploited in the past and so the younger age classes will be more common. However, in some Republics such as Estonia and White Russia, Group II forests have almost as large a concentration of mature and overmature timber as is found in Group III forests.

The Location of the Logging Industry

From the argument above it would seem that there is unlikely to be any problem in maintaining and increasing the output of timber from the Soviet Union's forests in the future. The present rate of cutting represents about half a percent of the standing timber and about one-half of the annual increment - which could itself be increased by sylvicultural operations. Indeed, the allowable cut calculated by the Soviet authorities as a whole is 636.68 million metres per annum.

However, this is not the case as will be seen in the following chapters. Factors other than just timber availability limit the output from Soviet forests. One of these factors is location; if we look at the location of logging operations in 1970 and 1975 -

shown in detail in Table 2.10 of the Appendix - it is evident that the greatest percentage of logging output comes from the European area of the Soviet Union. Of the 385 million cubic metres logged in 1970, 236 million were felled in European Russia; this is about 61 per cent, extracted from an area which contains only about 18 per cent of the forested land. In 1975 the position was much the same: of the 395 million cubic metres felled in that year, European Russia accounted for 60 per cent.

Previous five-year plans have emphasised the exploitation of the Asian forests and the relocation of logging activities to that area. However, this seems to have been unsuccessful, as the figures in Table 2.10 show. The main reasons for this have been the problems of labour recruitment and underfulfilment of investment plans; this has caused a slight re-emphasis in the plans. The eleventh five-year plan called for 'fuller use of the forest resource in the European part of the country' (Kirin, 1981). This means an increase in the use of smaller, younger timber and an increase in the output of timber from thinning operations; this is beginning to come into effect. In the 1960s thinnings accounted for a negligible proportion of the total output. In the later 1970s, however, over 10 per cent of output was made up of thinnings. In the Baltic States, for example, thinnings represented over half of the timber extracted.

The location of the forest products industries shows a similar pattern to that of logging: about 60 per cent of the country's sawn timber in 1975 was produced in the European part of the Soviet Union, as were 90 per cent of the particle board, 82 per cent of the fibreboard and 93 per cent of the plywood.

The general question of location will be discussed later, but now it might be useful to look at specific areas to describe the importance of logging and the forest products industries to the various regions of the Soviet Union.

THE RUSSIAN SOVIET FEDERATION OF SOCIALIST REPUBLICS

The RSFSR is the largest and most important Republic in the Soviet Union. It occupies 76.3 per cent of the area of the country and has a population of about 135 million. The area is divided into several constituent economic regions.

The Central Region

This is the area around Moscow. It is a region that has had a long history of over-exploitation of its forest resource. Now the region, which constitutes 2.2 per cent of the total area of the Soviet Union, contains only 1.7 per cent of the country's forests despite the fact that it is situated within the southern limits of the taiga and the northern limits of the mixed forest subzone. The exploitation of the forests in the past is indicated by their age class distribution. Of the overall growing stock of 1,542.5 million cubic metres, only 333.3 million, or 22 per cent, are mature or overmature. In the region 36.1 per cent of the forest area falls into Group I and 53.6 per cent is classified as Group II. Only 10.3 per cent comes into Group III.

The forests in this region are still over-exploited in certain ways. Although only about 85 per cent of the allowable cut is usually felled, the allowable cut for coniferous timber is over-exploited. In 1975, for example, 103.5 per cent of the coniferous allowable cut was felled; in some parts of the region up to one-third of output was in the form of intermediate fellings - indicative of a more intensive approach to sylviculture.

Within this region about 7 per cent of all of the Soviet Union's timber is felled, over 8 per cent of all the sawn timber is produced along with 16 per cent of the plywood and over 3 per cent of the paper. About one-third of the volume logged here comes from the area around Kostroma in the north-east; however, an extra 6 million cubic metres has to be imported into the region. Most comes from the North-west economic region, although a significant amount also arrives from Western Siberia and the Urals.

The Volga-Vyatka Region

Again this is an area situated in European Russia, to the east of the Central Region. It straddles three forest areas, the southern taiga, the mixed forest belt and the forest steppe. The total forest area in the region under the control of the forestry service is 11,682.6 thousand hectares and the growing stock is 1,209 million cubic metres. This area would appear to have a greater proportion of mature and overmature timber than the Central Region - about 44 per cent; however, again the forests are mainly in Groups I and II.

The forest products industries in this region are very important. The major centre is in the north of the region around the town of Kirov, although there is some form of forest industry present in almost every part of the region. The region produces about 7 per cent of the country's sawn timber, 4.2 per cent of the plywood, but more importantly, 12 per cent of the ordinary paper and 5 per cent of the kraft paper and cardboard. The paper industry is very important in this region with one factory in the area producing about one-third of the country's newsprint. In some of the concerns in the region there is almost total utilisation of the raw material with nearly all of the waste being used in the production of some other product or for heating.

The region imports 2.5 million cubic metres of timber from other regions, mostly in the form of pulpwood for its paper industries. Much of this comes from the Central Region and some from the Urals and the North-west Region; the region also exports timber, about four times as much as it imports, some of which goes to the Central Region and the Ukraine.

The Central Black-earth Region

This region lies to the south of the Central Region and centres itself around Voronezh. It covers three vegetation subzones - the mixed forest, the forest steppe and the Steppe. In the past the region was heavily forested, especially in the north, but now only about 10 per cent of the region is forest covered. Most of these forests are now concentrated in strips running north-south from Lipetsk to Voronezh and around Tambov. The area of forest is 1,321.3 thousand hectares and the growing stock is 150.4 million cubic metres.

Not surprisingly, only about 11 per cent of this is classified as mature or overmature. There are no Group III forests in the region, and most of the forests are in Group I. The area is very favourable for forest growth with some of the mature stands having growing stocks of about 300 cubic metres per hectare. The operations in the region's forests are such that thinnings play an important part and in some areas they account for over half of the timber felled.

Sixty per cent of the 600 thousand cubic metres harvested in the area is felled in the north of the region and a high proportion of it is firewood, although raw material for plywood manufacture is important. About 1.5 million cubic metres of timber are imported into the region, much of it as pit props for the region's industry.

The Povolzhski Region

The area of forests under the control of the forestry services in this region is 10,605.2 thousand hectares, and the growing stock is 1,023.4 million cubic metres of which about 38 per cent are classified as mature and overmature. Most of these forests are in the Bashkir ASSR and in Ufa there is an extensive woodworking industry and wood chemical plants.

Group I forests occupy 40 per cent of the total forest area of the region and most of these are agricultural shelterbelts, etc. Group II forests make up 33 per cent and the remaining forests are included in Group III; much of the forest here is made up of oak but most of it is of a very poor quality. The allowable cut is not fully taken up in this region - even about one-tenth of the allowable cut of the coniferous species is left standing each year. The region accounts for only 3 per cent of the timber felled in the Soviet Union but for about 6 per cent of the sawn timber, 11 per cent of the plywood and 7 per cent of the paper production. Much of the timber needed in the region is imported - about 8.2 million cubic metres in all; this mainly comes from the Volga-Vyatka, Urals and Western Siberia economic regions.

The Northern Caucasus Region

This economic region is located north of the other Republics in the Caucasus area. The major centre of the region is Rostov-on-Don. It is an important industrial region with coal, oil and chemical industries. It is only 10 per cent forest covered - an area of 3,460.9 thousand hectares with a growing stock of 391.1 million cubic metres. Most of these forests are in the area around Krasnodar towards the south of the region. Most (69 per cent) of the forests are classified as Group I and the rest as Group II. Oak is the main species in the area with pine being the commonest coniferous tree. The area over-cuts its coniferous growing stock by 14 per cent, although its broadleaved allowable cut is left under-utilised so that, overall, 99 per cent of the total allowable cut is felled.

Almost 1 per cent of the timber felled in the Soviet Union comes from this region and the area accounts for 2.6 per cent of the output of sawn wood and about the same amount of the country's production of plywood. About one-third of all of the timber felled in the region comes from thinnings and the region imports more timber than it fells, much of which is in the form of pit props, sawn timber and construction timber. Most of this comes from the Urals and Siberia. Although much of this imported timber is needed for the area around Krasnodar there are important furniture factories in Rostov and in other large towns in the region.

The Urals Region

The Urals have a long history of timber use; the timber in the Urals was converted into charcoal for use in the iron and steel industry before the Revolution. The earliest mechanisation of logging operations occurred in this region. Even today 33 per cent of the region is forest covered - 22,319.6 thousand hectares of forest coming under the control of the forestry authorities with a growing stock of 3,014.5 million cubic metres.

From the forestry point of view, the region divides itself into two areas. In the south the early exploitation of the forests has left the area with less forest than the areas in the north around Sverdlovsk and Perm; here are the major centres of forestry in the region. Most of the region's 43 per cent of forests which are included in Group III are to be found in this northern area. The major species are the valuable spruces, pines and birch.

The region accounts for almost 15 per cent of the total volume of timber felled in the Soviet Union making it the third most important region in terms of timber harvested after the North-west and Eastern Siberia. About 11 per cent of the sawn timber, 11 per cent of the plywood and about 22 per cent of the country's paper are produced in this region with paper making combines at Solikansk, Krasnokamisk, Perm and Krasnovishersk. All of these are to be found in the region of Perm as it is here that spruce, suitable for paper making, is most common. Around Sverdlovsk pine is the more common tree and so this area concentrates on sawmilling. As well as these industries there are chemical industries using the waste of wood processing, a healthy industry producing prefabricated housing units from wood as well as the usual plywood and particle board plants.

Although the region imports 1.7 million cubic metres of timber, almost entirely from Siberia, it is a net exporter of timber, exporting in total about 15 million cubic metres, most from the Perm *oblast*. Most of the exported timber goes to the Ukraine, the Povolzhski Region and the Northern Caucasus and most of the exports are in the form of sawn timber and pit props.

The North-west Region

This region is situated in the north-west of the European part of the RSFSR and contains the important forested areas of Karelia, Murmansk, Archangelsk and the Komi ASSR. The area sits squarely in the taiga and is well endowed with forests; about 64 per cent of the region is forest covered - 105,943.4 thousand hectares of forest with a growing stock of 7,366.5 million cubic metres are under the control of the forestry authorities. Almost one-third of the forests are situated in the Komi ASSR in the east of the region and most of the rest is in Karelia and in the north around Murmansk and Archangelsk; by volume, 72 per cent of these forests are mature or overmature.

However, the region is undergoing a period of transition. In the south, around Leningrad, Pskov and Novgorod, there is a long history of over-exploitation; here all of the forests are either in Group I or II and the ages of the timber illustrate this. In the Leningrad area the average age of coniferous timber is fifty years, whilst in the Komi ASSR it is 120. Also the importance of thinnings in the various areas differs; around Leningrad they account for as much as 17 per cent of output whilst in the Komi ASSR they hardly contribute to output at all. In the northern areas the forests are being over-exploited and the allowable cut for coniferous species which make up most of the forests is being exceeded by about 30 per cent.

The region is a significant one for the production of timber and forest products. About one-quarter of the timber felled in the Soviet Union is harvested in this region and about 16 per cent of all of the sawn timber is produced here as is 19 per cent of the plywood and 40 per cent of the paper. The most important area in this context is the Archangelsk one. The logging operations have been concentrated in the basins of the Northern Dvina, Onega and other rivers as well as along the railway lines from Vologda to Archangelsk, Konosha to Ukhta and Kem to Petrozavodsk.

The industries in the region are usually in the form of pulp and paper plants or sawmills which are sometimes combined with units for the production of panel products, furniture and packaging. Large complexes are located in Sitvitkar, the capital of the Komi ASSR, although almost every town of significance in the area produces forest products. Much of the output of the area's forests is exported from the country from the ports of Archangelsk and Murmansk. Output from the Komi ASSR is usually directed inwards towards the domestic economy; the region imports little timber from the rest of the country, about 1 million cubic metres per annum, mostly from the Volga-Vyatka and Western Siberia economic regions. More importantly, apparently because of a lack of processing capacity, about 26 million cubic metres of timber are exported annually to other parts of the Soviet Union. Most of this goes to the Ukraine, White Russia and the Central Region of the RSFSR.

The Western Siberia Region

About 60 per cent of this large region is forested. The region east of the Urals but the western-most of the Asiatic regions of the RSFSR, contains 138,603 thousand hectares of forest with a growing stock of 9,654.8 million cubic metres; of this, 7,002 million, or 73 per cent, is mature or overmature. Pine and birch are the commonest species in the forests of the region with larch beginning to put in an appearance in the east. About two-thirds of the region's forests are in the Tumen *oblast* in the west of the region. In the west, especially in the Tumen *oblast* little of the forest is classified as mountain forest but as one moves further east about 70 per cent of the forest is considered to be 'mountainous'.

The region is as yet under-utilised from the point of view of forestry. Only one-third of the allowable cut is in fact harvested, although for the coniferous species alone the figure is a little higher at 44 per cent. The region contributes about 8.5 per cent of the total volume felled in the Soviet Union, produces 7.6 per cent of the sawn timber, 3 per cent of the plywood and almost 2 per cent of the cardboard. The main centres of logging and woodworking are the south and central taiga belts. The towns of Tumen, Tomsk and Novosibirsk are important for sawn timber and Tomsk is also important for the production of plywood, pencils and matches. About 6.5 million cubic metres of timber are exported from the region, most of it in the form of either sawn timber or pit props; most of this comes from the Tumen *oblast* and goes to Central Asia, the Urals and the Ukraine.

The Eastern Siberia Region

This region (70 per cent forest covered) encompasses 286,058.5 thousand hectares of forest. The growing stock is 27,191 million cubic metres, of which 20,185.8 (74 per cent) is mature or overmature. About 85 per cent of the region's forests are coniferous with larch being the most common species. In this area trees reach only 20 centimetres breast height diameter after about 120 years. One of the major problems in the exploitation of the forests in this region is the absence of a developed transport system; on average, every 100 hectares of forest in Eastern Siberia contains only 20 metres of road and most of the roads are only dirt tracks which will only be usable for part of the year. As a result, only about 30 per cent of the forest area is deemed to be accessible and the allowable cut is only 36 per cent utilised - 48 per cent for coniferous species and only 6 per cent for broadleaved species. Only one-quarter of the current annual increment is extracted from these forests.

The region contains Lake Baikal and the Baikal-Amur Magistral or railway (BAM); the first is important as a centre for timber conversion and some of the forests around the lake are used for the protection of the lake and environmental conservation - not very successfully it would appear. BAM is important in an area with such a poor communication system and should increase the accessibility of the forests in the area; more of this will be outlined in Chapter 3.

The region accounts for 17.5 per cent of the country's timber output and for 15 per cent of the sawn timber, 13 per cent of the paper and 2 per cent of the plywood produced, and the main areas of production are around the railways. There are centres for the mechanical processing of timber at Bratsk, Chita, Krasnoyarsk and Igarka on the lower reaches of the Yenesei and at many other smaller towns in the south of the region. Also there are important pulp and paper producing units in Bratsk, Krasnoyarsk and on Lake Baikal.

Because of the lack of capacity of the processing industries in the region, much of the timber felled in the region is either transported to the south of the region or exported out of the area. About 12 million cubic metres of timber leave the region in this way, mostly to central Asia, the Ukraine, the Povolzhski Region and for export from the country altogether.

The Far East Region

This eastern-most region of the Soviet Union contains 285,123.4 thousand hectares of forest with a total growing stock of 24,900 million cubic metres. Coniferous species account for 84 per cent of the timber in the region and larch covers half of the area there with pine and cedar the next most common species. The volume logged in the region is increasing slowly from 12.6 million cubic metres in 1950 to 28.6 million in 1975, an increase of just over 3 per cent per annum. The output represents just over 40 per cent of the allowable cut, which is itself fairly low because of the lack of accessibility of many of the forests in the region. About 5 per cent of the timber felled in the Soviet Union comes from the Far East and it exports about 9.2 million cubic metres of timber, some of it abroad.

The area is undergoing development; twenty logging concerns have been built in the Khabarovsk area over the last ten years, along with more than thirty units for the production of chips from low quality wood. However, at the same time the pulp and paper industry in the region has only been developing slowly.

OTHER REPUBLICS OF THE SOVIET UNION

The Ukraine, White Russia and Moldavia

Together these European Republics are all in the lightly forested areas of the Soviet Union but contain 17,237.2 thousand hectares of forest with a growing stock of 1,357.6 million cubic metres. All of the forests are in either Group I or II and the allowable cut is almost entirely taken up. The three Republics together import timber from the rest of the country. The Ukraine imports 18.6 million cubic metres annually, including one million cubic metres of firewood; White Russia imports 1.85 million and Moldavia 1.3 million cubic metres.

The Baltic States

The Baltic States consist of Estonia, Latvia and Lithuania and represent an important industrial area with a developed woodworking industry. The area encompasses 4,296.7 thousand hectares of forest containing about 506 million cubic metres of timber. The forests are old with more than 70 per cent of them being mature or overmature and all of the forests are in either Group I (43 per cent) or Group II. The allowable cut of about 3 million cubic metres per annum is almost fully utilised and almost this much again is imported into the region, mostly from the North-west economic region. Of the Republics, Latvia and Lithuania are the most important from the point of view of production of forest products with woodworking plants in Vilnus and Riga.

The Caucasus

This region consists of Georgia, Armenia and Azerbaijan and the total forest area is 4,184.4 thousand hectares, representing about a quarter of the area of the whole region. Again there are no Group III forests in the area but the forests there are of a relatively high productivity. Total output from the area is about 660 thousand cubic metres, a considerable proportion of which is firewood; most of this comes from Georgia. About 2 million cubic metres of timber have to be imported into the region to satisfy the industry there is, most of which is located in the east of Georgia.

The Central Asian Republics

Finally we come to the Central Asian Republics of Kazakhstan, Kirghizia, Tadzhikistan, Turkmenistan and Uzbekistan. Together the forest area of these Republics is 26,594.6 thousand hectares - about 6 per cent of their combined area. All of the forests in these lightly forested areas are either Group I or Group II except for some in Kazakhstan and the combined growing stock of these forests is 317.2 million cubic metres. Much of the forest in this region has a protective significance and only about 2 million cubic metres are felled annually there. There are no really significant centres for wood processing industries in the region.

SOME BRIEF CONCLUSIONS

As a brief conclusion to this chapter we can say that it would appear possible, at first sight at least, to double the timber output from the Soviet Union's forests in the relatively short-term. The timber felled annually in the country as a whole is less than half of the allowable cut which is itself less than the gross current annual volume increment. In the longer term, the geographical distribution of the timber-using industries could be much improved. Much timber has to be transferred from the North-west, Siberia and the Far East to the lightly forested but industrially developed areas of central Europe. This industrial distribution, in part, has led to the over-cutting of forests in some parts of the country and an under-utilisation of them in other areas.

In the rest of the book we will examine reasons why the forest industries cannot simply follow the policies of expansion and relocation and why factors other than biological ones will be effective in limiting the output of timber from the Soviet Union's forests and limiting it at a level far below the allowable cut.

APPENDIX
TO CHAPTER 2

TABLE 2.1 The Distribution of the Major Coniferous Species
(All figures are in thousands of hectares)

	Pine	Spruce	Fir	Larch
Soviet Union Total	141,611	72,105	23,100	274,261
RSFSR*	135,230	70,126	22,443	273,953
North	13,390	28,723	49	376
North-west	9,203	4,909	0	0
Centre	6,564	4,807	9	5
Volga	400	26	1	0
Urals	5,587	8,235	527	59
North Caucasus	160	13	124	0
West Siberia	33,349	4,750	4,234	4,807
East Siberia	61,776	4,462	16,576	215,432
Far East	4,801	14,201	923	53,274
Armenia	1	0	0	0
Azerbaijan	0	0	0	0
Georgia	74	119	200	0
Estonia	336	146	0	0
Latvia	736	229	0	0
Lithuania	465	274	0	0
Moldavia	0	0	0	0
Ukraine	1,645	532	74	1
White Russia	2,515	415	0	2
Kazakhstan	609	160	379	305
Kirghizia	0	104	4	0
Tadzhikistan	0	0	0	0
Turkmenistan	0	0	0	0
Uzbekistan	0	0	0	0

* These figures are taken from Tseplayev (1965) and the division of the RSFSR used is not exactly the same as the division into economic regions described in the text.

TABLE 2.2 The Distribution of the Major Broadleaved Species
(All figures are in thousands of hectares)

	Birch	Aspen	Beech	Oak
Soviet Union Total	91,726	14,947	2,465	8,511
RSFSR*	89,599	13,895	728	6,367
North	7,913	867	0	0
North-west	3,140	514	0	17
Centre	7,650	3,023	0	1,572
Volga	296	573	0	1,062
Urals	7,157	1,591	0	649
North Caucasus	183	28	728	942
West Siberia	20,166	4,263	0	0
East Siberia	26,899	2,688	0	0
Far East	16,195	348	0	2,125
Armenia	1	0	85	83
Azerbaijan	1	3	248	245
Georgia	0	8	961	158
Estonia	205	20	0	1
Latvia	301	60	0	2
Lithuania	138	70	0	15
Moldavia	0	1	0	122
Ukraine	262	71	443	1,306
White Russia	602	154	0	209
Kazakhstan	611	215	0	3
Kirghizia	4	0	0	0
Tadzhikistan	2	0	0	0
Turkmenistan	0	0	0	0
Uzbekistan	0	0	0	0

* These figures are taken from Tseplayev (1965) and the division of the RSFSR used is not exactly the same as the division into economic regions described in the text.

TABLE 2.3 *Goslesfond* (All figures are in thousands of hectares)

	Total Area of Goslesfond	Forested Area		Unforested Area
		Total	Forest Covered	
Soviet Union total	1,229,657.4	933,820.6	768,774.1	295,836.8
European part	236,905.5	197,440.6	184,338.9	39,464.9
Asian part	992,751.9	736,380.0	584,435.2	256,371.9
RSFSR	1,161,423.1	879.633.6	729,705.5	281,789.5
European part	206,126.4	169,848.2	158,842.8	36,278.2
Asian part	955,296.7	709,785.4	570,862.7	245,511.3
Armenia	442.6	375.3	295.6	67.3
Azerbaijan	1,194.0	1,079.4	950.8	114.6
Georgia	2,983.3	2,729.7	2,356.8	253.6
Estonia	2,254.1	1,874.1	1,763.6	380.0
Latvia	3,062.0	2,520.0	2,457.6	542.0
Lithuania	2,032.8	1,776.7	1,723.3	256.1
Moldavia	343.7	327.5	271.2	16.2
Ukraine	10,241.2	9,268.3	8,614.2	972.9
White Russia	8,225.4	7,641.4	7,063.0	584.0
Kazakhstan	21,762.9	16,583.3	9,109.2	5,179.6
Kirghizia	2,651.4	1,203.7	790.9	1,447.7
Tadzhikistan	1,774.1	812.5	407.0	961.6
Turkmenistan	6,074.4	4,282.5	2,074.0	1,791.9
Uzbekistan	5,192.4	3,712.6	1,191.4	1,479.8

Source: Vorobiev *et al.* (1979)

TABLE 2.4 The Forested Area of *Goslesfond* (All figures are in thousands of hectares)

	Total	Forest Covered		Non-forest Covered		
		Total	Under Very Young Plantations	Sparsely Wooded	Burnt Over and Dead Areas	Clear Felled in Recent Years
Soviet Union total	933,820.6	768,774.1	7,912.7	89,106.8	54,316.0	13,711.0
European part	197,440.6	184,338.9	4,620.2	815.7	1,891.3	5,774.5
Asian part	736,380.0	584,435.2	3,292.5	88,291.1	52,424.7	7,936.5
RSFSR	879,633.6	729,705.5	4,509.8	78,782.0	54,113.2	12,523.1
European part	169,848.2	158,842.8	3,322.4	415.3	1,868.8	5,398.9
Asian part	709,785.4	570,862.7	1,187.4	78,366.7	52,244.4	7,124.2
Armenia	375.3	295.6	53.1	26.6	0	0
Azerbaijan	1,079.4	950.8	51.9	76.7	0	0
Georgia	2,729.7	2,356.8	112.8	260.1	0	0
Estonia	1,874.1	1,763.6	62.5	15.6	1.1	31.3
Latvia	2,520.0	2,457.6	36.9	0.6	0.7	24.2
Lithuania	1,776.7	1,723.3	38.7	0.9	0.9	12.9
Moldavia	327.5	271.2	37.4	0.3	0.2	18.4
Ukraine	9,268.3	8,614.2	536.4	13.0	13.1	91.6
White Russia	7,641.4	7,063.0	368.1	6.6	6.5	197.2
Kazakhstan	16,583.3	9,109.2	1,131.2	5,534.9	161.6	646.4
Kirghizia	1,203.7	790.9	128.3	252.4	2.2	29.9
Tadzhikistan	812.5	407.0	80.0	323.7	1.8	0
Turkmenistan	4,282.5	2,074.0	389.4	1,812.4	6.7	0
Uzbekistan	3,712.6	1,191.4	376.2	2,001.0	8.0	136.0

Source: Vorobiev *et al.* (1979)

TABLE 2.5 The Unforested Area of *Goslesfond* (All figures are in thousands of hectares)

	Total	Pastures and Arable Land*	Water	Roads etc.	Unused land
Soviet Union total	295,836.8	16,220.9	20,459.1	4,261.3	254,895.5
European part	39,464.9	2,832.1	4,000.7	2,185.9	30,446.2
Asian part	256,371.9	13,388.8	16,458.4	2,075.4	224,449.2
RSFSR	281,789.5	10,153.9	20,102.8	3,546.6	247,986.2
European part	36,278.2	2,049.6	3,894.3	1,639.7	28,694.6
Asian part	245,511.3	8,104.3	16,208.5	1,906.9	219,291.6
Armenia	67.3	27.0	0.3	0.3	39.7
Azerbaijan	114.6	57.0	0.6	11.2	45.8
Georgia	253.6	126.0	1.5	12.0	114.1
Estonia	380.0	37.3	9.4	37.3	296.0
Latvia	542.0	118.8	15.2	88.3	319.7
Lithuania	256.1	86.4	1.1	69.9	98.7
Moldavia	16.2	0.6	0.2	0.2	15.2
Ukraine	952.9	217.4	62.1	207.0	486.4
White Russia	584.0	112.0	16.0	120.0	336.0
Kazakhstan	5,179.6	2,589.8	152.3	130.6	2,306.9
Kirghizia	1,447.7	695.8	18.3	18.2	715.4
Tadzhikistan	961.6	201.5	39.6	0.9	719.6
Turkmenistan	1,791.9	1,109.6	3.2	3.2	675.6
Uzbekistan	1,479.8	687.8	36.5	15.6	739.9

* Pastures here contain the categories *pashini*, *syenokosi* and *pastbishcha* - arable land, areas for hay-making and pasture.

Source: Vorobiev *et al.* (1979)

TABLE 2.6 Percentage of Forested Area by Species(All figures are percentages)

	Coniferous Species				Broadleaved Species		Other Species
	Total	Pine	Spruce/ fir	Larch	Total	alder/aspen birch	
Soviet Union total	75.1	17.0	13.7	38.5	19.8	16.5	5.1
European part	63.0	30.1	32.2	0.3	36.6	29.0	0.4
Asian part	78.4	13.1	8.5	49.1	15.1	13.0	6.5
RSFSR	76.2	16.5	13.8	39.9	18.3	16.4	5.5
European part	64.8	28.9	35.1	0.3	32.7	30.4	2.5
Asian part	79.1	13.3	8.5	49.7	14.6	13.0	6.3
Armenia	2.8	1.4	-	-	93.0	2.5	4.2
Azerbaijan	0.1	0.1	-	-	90.9	2.8	9.0
Georgia	18.7	3.7	15.0	-	73.9	7.5	7.4
Estonia	68.4	47.7	20.6	0.1	31.5	31.1	0.1
Latvia	67.5	50.9	16.5	0.1	32.5	31.8	-
Lithuania	63.1	45.1	18.0	0.1	36.9	34.3	-
Moldavia	0.2	0.2	-	-	97.0	4.7	2.8
Ukraine	47.3	35.8	11.4	0.1	52.0	10.7	0.7
White Russia	67.7	58.2	9.4	0.1	32.3	27.8	-
Kazakhstan	30.3	14.9	11.2	3.3	52.9	22.6	16.8
Kirghizia	34.3	0.4	17.8	0.2	8.0	2.0	57.7
Tadzhikistan	31.9	-	-	-	20.7	3.2	47.4
Turkmenistan	26.4	-	-	-	16.9	0.2	56.7
Uzbekistan	9.6	-	-	-	56.3	0.4	34.1

Source: Vorobiev *et al.* (1979)

TABLE 2.7 The Management Division of Forests

(All figures are percentages)

	Soviet Union	European part	Asian part
GROUP I	15.8	4.9	10.9
green belts	1.5	1.1	0.4
shelter belts	1.8	0.7	1.1
belts around lakes, along rivers etc.	5.8	1.3	4.5
protective belts along roads and railways	0.4	0.3	0.1
pretundra belt	4.3	1.0	3.3
other	2.0	0.5	1.5
GROUP II	6.9	4.9	2.0
exploitable forests	6.2	4.7	1.5
special zones and belts for protection along rivers etc.	0.3	0.1	0.2
other	0.4	0.1	0.3
GROUP III	77.3	8.4	68.9
exploitable forests	34.7	8.1	26.6
reserve forests	32.4	0.2	32.2
protective belts etc.	0.9	0.1	0.8
other	9.3	0	9.3

Source: Vorobiev *et al.* (1979)

TABLE 2.8 The Annual Increment in Soviet Forests

	Total*	Coniferous species*	Broad-leaved species*	Average per hectare of forest-covered area
Soviet Union total	881.06	600.16	280.90	1.15
European part	291.71	150.95	140.76	1.58
Asian part	859.35	449.21	140.14	1.01
RSFSR	821.08	568.19	252.82	1.13
European part	237.58	121.84	115.74	1.50
Asian part	583.50	446.35	137.15	1.02
Armenia	0.31	0.01	0.30	1.05
Azerbaijan	1.34	0.00	1.34	1.41
Georgia	3.53	0.86	2.67	1.50
Estonia	2.56	1.60	0.96	1.45
Latvia	4.94	2.90	2.04	2.01
Lithuania	3.77	2.15	1.62	2.19
Moldavia	0.59	0.00	0.59	2.18
Ukraine	21.49	11.76	9.73	2.49
White Russia	15.60	9.83	5.77	2.21
Kazakhstan	5.53	2.72	2.81	0.61
Kirghizia	0.13	0.10	0.03	0.16
Tadzhikistan	0.03	0.01	0.02	0.07
Turkmenistan	0.01	0.00	0.01	0.01
Uzbekistan	0.15	0.03	0.12	0.13

* millions of cubic metres

Source: Vorobiev *et al.* (1979)

TABLE 2.9 The Distribution of Mature and Overmature Timber by Area (All figures are percentages)

	All species		Coniferous species	
	Group III	Group II	Group III	Group II
Soviet Union total	85.94	4.54	87.85	3.39
European part	68.21	16.18	75.86	11.85
Asian part	89.70	2.08	90.07	1.82
RSFSR	86.68	4.29	88.24	3.24
European part	68.21	15.32	77.52	11.15
Asian part	89.95	2.04	90.19	1.80
Armenia	-	-	-	-
Azerbaijan	-	-	-	-
Georgia	-	0.81	-	-
Estonia	-	79.97	-	79.00
Latvia	-	72.15	-	72.70
Lithuania	-	72.26	-	61.45
Moldavia	-	21.39	-	-
Ukraine	-	60.54	-	66.62
White Russia	-	76.90	-	86.39
Kazakhstan	-	14.41	-	10.76
Kirghizia	-	-	-	-
Tadzhikistan	-	-	-	-
Turkmenistan	-	-	-	-
Uzbekistan	-	15.52	-	-

Source: Vorobiev *et al.* (1979)

TABLE 2.10 The Distribution of Output of Timber and Forest Products

	Timber Felled million m^3		Sawn Wood million m^3		Particle-board thousand m^3		Fibreboard million m^2		Plywood thousand m^3	
	1970	1975	1970	1975	1970	1975	1970	1975	1970	1975
Soviet Union total	385.0	395.0	116.4	116.2	1,994	3,994	208,3	408.0	2,045	2,196
European part	257.5	254.1	80.3	69.6	1,811	3,585	192.9	334.0	1,890	2,046
Asian part	127.5	140.9	36.1	42.4	183	409	15.4	74.0	155	150
RSFSR	361.6	374.6	99.2	87.3	1,376	2,274	170.9	324.0	1,583	1,687
European part	236.0	235.7	66.1	48.9	1,229	1,937	155.5	250.0	1,428	1,537
Asian part	125.6	138.9	34.9	38.4	147	337	15.4	74.0	155	150
Caucasus[1]	0.9	0.8	1.2	4.1	13	361	-	-	6	*
Baltic States[2]	8.5	6.0	2.7	3.2	112	225	16.0	30.0	46	*
Moldavia	0.2	0.2	0.2	0.2	9	10	-	-	2	*
Ukraine	6.7	6.2	7.8	10.0	345	796	4.5	25.0	195	*
White Russia	5.2	5.2	2.3	3.2	103	256	16.9	29.0	213	*
Central Asian States[3]	1.9	2.0	1.2	4.0	36	72	0	0	0	

Notes: 1 - Armenia, Azerbaijan and Georgia
2 - Estonia, Latvia and Lithuania
3 - Kazakhstan, Kirghizia, Tadzhikistan, Turkmenistan and Uzbekistan

* Figures not available in a consistent series

Source: Glotov (1977) and Stupnev *et al.* (1976)

3
The Industrial Forests of the Soviet Union

INTRODUCTION

The main sources of timber for the Soviet economy are the forests of Groups II and III and the division of forests into groups was described in the last chapter. Briefly, Group III forests are to be found in the heavily forested regions of the Soviet Union, that is the northern European parts of Russia, Siberia and the Far East and Group II forests are in the lightly forested parts of the country. At the present moment about two-thirds of the logging operations, whether they are carried out in Group II or Group III forests, are under the control of the recently re-organised Ministry of Timber, Woodworking, Pulp and Paper Industries - *Minlesbumprom*. Most of the remaining logging operations are under the control of the State Committee for Forestry of the Council of Ministers - *Gosleskhoz*. This will be examined later in the chapter.

In the main part of this chapter we will look at the organisation of *Minlesbumprom* and follow this by a discussion of the various aspects of the logging process which are of interest. Whilst most of the evidence and points of view presented in this second section of the chapter relate to the operations of *Minlesbumprom*, they are of general significance for logging operations wherever they occur in the Soviet Union.

THE GENERAL ORGANISATION OF *MINLESBUMPROM*

Minlesbumprom was formed at the beginning of 1981 by a merger of the timber industries' Ministry and the Ministry for Paper Production. Looking at the timber industries' section of the Ministry there are more than 3,000 enterprises and organisations situated in most of the parts of the country, including those areas with no Group III forests. The organisations and enterprises in such lightly forested parts of the country are concerned mainly with the woodworking side of the forest industries and the logging and timber production is left to *Gosleskhoz*.

In 1971 the basic organisational structure concerning the various enterprises under the control of the then Ministry for the Timber Industries was altered (*Minlesprom*, 1971); until 1971 there had been five levels of management in the forest industries. First came the Ministry, then a number of *glavki* each dealing with one line of production. This was followed by a layer of *obyedineniye* which were groupings of enterprises working in a specific geographical region, and underneath this came a number of smaller combines and finally came the individual enterprises themselves. However, it was found that such a system had serious drawbacks and was overburdened by administrative apparatus; the main problems arose in the areas of information and planning. Plans that came from the Ministry had to be broken down four times before the final enterprise plans were arrived at, and this process resulted in a considerable loss of information by the time that the directives from the Ministry had been translated into objectives for individual enterprises. Similarly, the plans produced by the individual enterprises as the first part of the planning process were aggregated many times before the figures reached the Ministry, with a resulting loss of specificity of the indicators that the Ministry could work with. Also, the fact that information travelled 'up' and 'down' the hierarchy of organisational structures and not from 'side' to 'side' led to incongruities. For example, enterprises producing particle board came under one *glavk* whilst sawmills came under the jurisdiction of another. This inhibited the use of the waste from sawmills in the production of particle board, when the two factories might be quite near to each other. As well as this it was observed that there were efforts to hide the true productive capacity of the enterprises coming under the control of a *glavk*. This was a reaction to the rather assymetric reward pattern for plan fulfilment observed in the Soviet Union. Basically, there was, and still is, a large discrepancy between the reward enjoyed by an organisation fulfilling a plan by 101 per cent and one 'enjoyed' by an organisation producing only 99 per cent of the planned total. Thus, the hiding of the productive capacity meant that enterprises might receive plans that were relatively easy to fulfil - that is plans which did not contain as targets the maximum output possible from an enterprise. Changes have been made in the system of rewards built into the planning process to try to overcome this problem, and these will be discussed in Chapter 7.

Starting in 1971 a re-organisation programme was instituted to change from a system of five levels of management to one of only three. Below the Ministry now come a number of *obyedineniye* and under these come the enterprises themselves. In some areas a four level system of management has been maintained with combines forming a layer of management above the individual enterprise but below the *obyedineniye*. This four-tier system is to be found in areas where combines formed an important management level prior to the reorganisation. Some of the former combines have been put directly under the control of the Ministry, effectively becoming *obyedineniye* in their own right. The present structure of the logging and woodworking side of *Minlesbumprom* is shown in Fig. 3.1.

Thirteen combines come under the direct control of the Ministry, such as *Bashles* and *Gorkles*. All of these combines with the exception of *Voronezhmebel* (a furniture-producing combine based in Voronezh) are concerned with logging and are regional organisations based in a specific area. There are also thirteen *obyedineniye* concerned with logging, e.g. *Archangelsklesprom*, and like the combines these are based in specific areas. As mentioned before, some

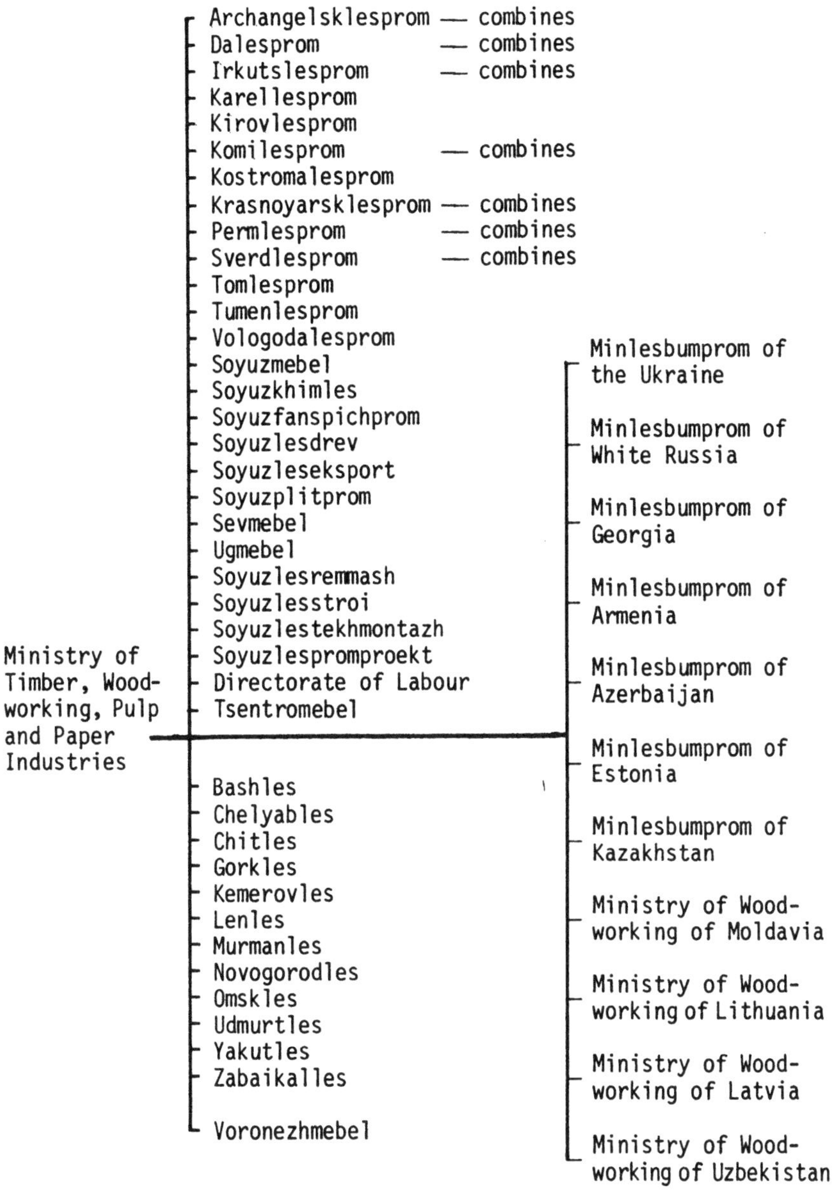

Source: Anon, 1974; Timofeyev, 1975

FIG. 3.1 The Organisation of the Forest Industries

contain combines and these are indicated in Fig. 3.1. Other of the *obyedineniye* are concerned with furniture manufacture: these are *Soyuzmebel*, *Ugmebel*, *Tsentromebel*, and *Sevmebel*. Some are concerned with panel products, *Soyuzplitprom*, *Soyuzlesdrev* and *Soyuzfanspichprom*; this is also responsible for the manufacture of matches, whilst others have more specialised responsibilities. For example, *Soyuzlesremmash* is the organisation that has the overall responsibility for major repairs and maintenance of the capital equipment in the logging enterprises of the Ministry, *Soyuzlestekhmontazh* and *Soyuzlesstroi* are concerned with the equipping and building of logging and woodworking units.

As well as these *obyedineniye* and combines there are also a number of Republic Ministries that come under the control of the main Ministry. These are usually located in the lightly forested areas of the country and are mainly concerned with the timber processing aspects of the forest industries.

The actual effects of this re-organisation are very difficult to gauge. No doubt the reduction of the number of tiers of management in the forest industries makes planning somewhat more simple. But such a re-organisation makes it less likely that local anomalies will be more swiftly dealt with. Information about such problems still has to travel to the 'top'. Indeed, reports are seen in the technical press of some logging concerns producing output for the woodworking industries far away whilst nearby woodworking plants are supplied from distant forests. The recent amalgamation of the former Ministry of Logging and the Timber Industries and the Ministry of Paper Production into one was partly inspired by the desire to strengthen the links between the producers and users of timber.

Lower Levels of Logging Organisation

There are twenty-five *obyedineniye* and combines concerned with logging under the control of *Minlesbumprom*. Some details of these organisations are given in Table 3.1 and the map in Fig. 3.2 shows the approximate areas that come under their jurisdiction. The table shows the species composition of the growing stock that are within the boundaries of the various enterprises. They are not really consistent with the figures in the tables in the Appendix to Chapter 2 as the figures in Table 3.1 refer only to the species in the area to be logged in the near future. From the data in this table, however, it is evident that pine and spruce are the main coniferous species to be found in the *obyedineniye* and combines in the west of the country and larch is very common in the east. However, larch is less common according to these figures than the overall figures in Table 2.6 suggest. The reason for this is probably that the logging concerns in the east of the country are concentrating on those areas where the more valuable pines and spruces grow. Of the broadleaved species, birch is by far the commonest; the small average tree sizes that confront Soviet foresters are also apparent from the table.

The management structure of logging is shown in Fig. 3.3. Within each *obyedineniye* or combine comes a series of 'forest industrial economies' or *lespromkhozi* and in all there are more than 600 of these. These are the basic planning units in the Soviet logging

TABLE 3.1 The *Obyedineniye* and Combines of *Minlesbumprom* (Species Composition %)

	Pinus spp	Picea spp	Larix spp	Cedrus spp	Abies spp†	Betula spp	Populus spp	Alnus, acer spp
1. Archangelsklesprom	20	70				10		
2. Dalesprom	10	20	60		10			
3. Irkutslesprom	50	10	20	*	10	10	*	
4. Karellesprom	50	40				10	*	
5. Kirovlesprom	10	50				20	20	
6. Komilesprom	30	50	*			20	*	
7. Kostromalesprom	*	50				30	20	
8. Krasnoyarsklesprom	20	20	10	*	30	10	10	
9. Permlesprom	*	50	*		20	20	10	
10. Sverdlesprom	30	30	*	*	10	30		
11. Tomlesprom	20	10	*	20	10	30	10	
12. Tumenlesprom	50	10	10	*		20	10	
13. Vologdalesprom	10	60				10	20	*
14. Bashles	10	10			10	30	20	20
15. Chelyables	10	20		*	10	30	30	
16. Chitles	30		50			20		
17. Gorkles	30	20				20	30	*
18. Kemerovles		*		*	40	30	30	
19. Lenles	10	40				20	30	
20. Murmanles	60	30				10		
21. Novgorodles	10	30				30	30	
22. Omskles	10	10		10	10	40	20	
23. Udmurtles	10	60			*	20	10	
24. Yakutles	30	10	60	*	*			
25. Zabaikalles	60	*	40	*	*			

The figures follow the Soviet practice of quoting the proportion of species to the nearest ten per cent.

* The species in question is prevalent but constitutes less than 5 per cent. † Includes *Pinus sibirica*

TABLE 3.1 continued

	Percentage of total output	Average tree volume	Average skidding distance	Percentage output of useful timber	Profitability		
					Loss making	10%	10-20%
1. Archangelsklesprom	9	0.3	26.5	78.5		*	
2. Dalesprom	8	0.5	38.0	70.0			*
3. Irkutslesprom	7	0.7	83.0	77.0			*
4. Karellesprom	6	0.3	33.0	78.0	*		
5. Kirovlesprom	7	0.3	33.0	72.4		*	
6. Komilesprom	7	0.2	33.0	77.0		*	
7. Kostromalesprom	3	0.3	31.8	65.8		*	
8. Krasnoyarsklesprom	8	0.6	41.8	76.7		*	
9. Permlesprom	8	0.3	28.6	79.5			*
10. Sverdlesprom	9	0.4	41.0	74.4		*	
11. Tomlesprom	3	0.5	36.0	67.7	*		
12. Tumenlesprom	3	0.4	37.0	77.3			*
13. Vologdalesprom	6	0.3	32.0	77.2			*
14. Bashles	1	0.3	32.0	65.8		*	
15. Chelyables	.75	0.2	36.0	75.1			*
16. Chitles	.5	0.4	88.0	78.8		*	
17. Gorkles	2	0.2	38.0	75.2			*
18. Kemerovles	1	0.4	43.6	69.7			*
19. Lenles	2	0.3	38.0	68.0			*
20. Murmanles	.5	0.2	27.0	81.6			
21. Novgorodles	1	0.3	37.0	64.1			*
22. Omskles	.5	0.3	47.0	70.5			
23. Udmurtles	1	0.3	44.0	70.3			*
24. Yakutles	.25	0.5	21.0	88.2			*
25. Zabaikalles	2	0.5	61.0	82.5			*

Source: Kozhin and Styazhkin (1976)

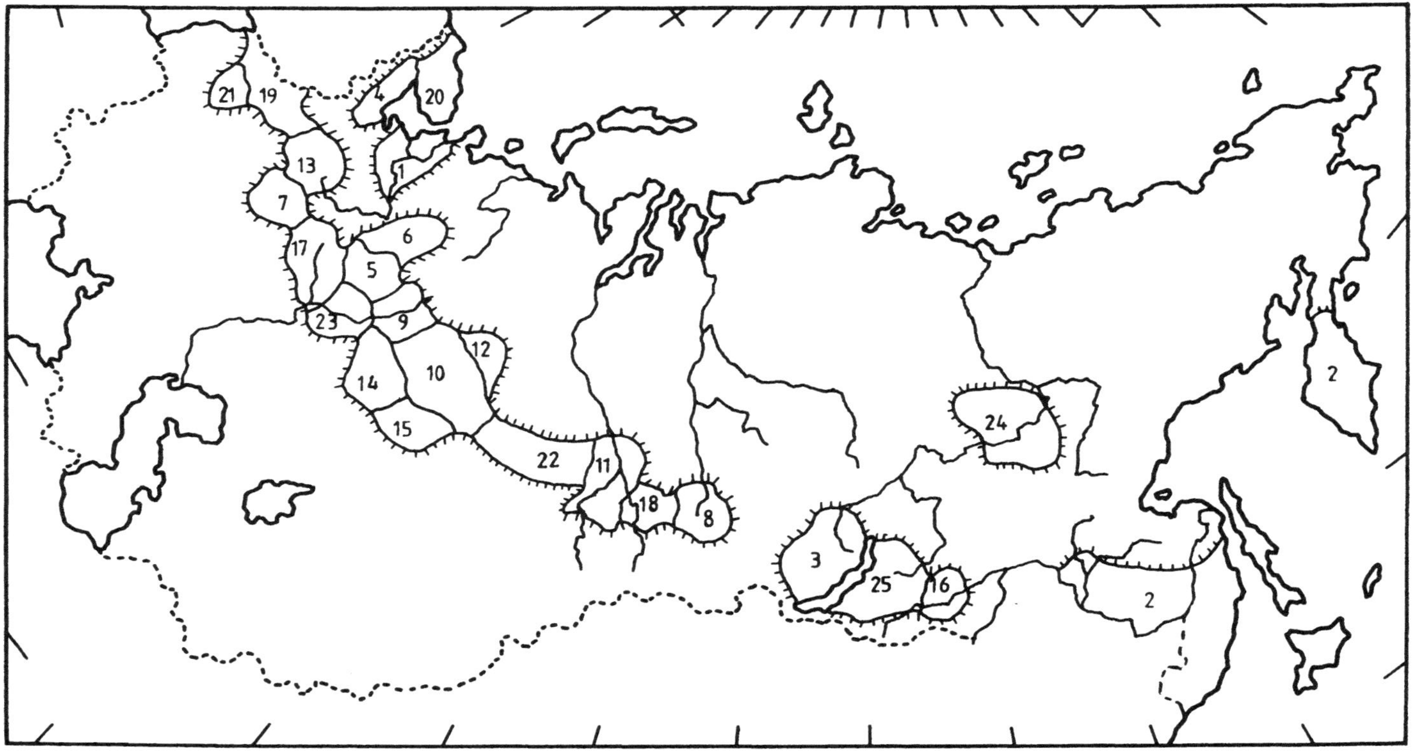

FIG. 3.2 The *Obyedineniye* and Combines of *Minlesbumprom* Figures refer to Table 3.1)

industry. The number of *lespromkhozi* that come under the control of the higher level of management varies from one part of the country to another, but usually it lies between fifteen and forty. The five-year plans that *Minlesbumprom* has to fulfil are broken down into one-year plans which are expressed at the *lespromkhoz* level. Each *lespromkhoz* is allocated an area of forest which is to be logged during the enterprise's operational life. The average annual output of a *lespromkhoz* is between 200 and 400 thousand cubic metres of timber although some lie outside this range.

The *lesopunkt* - the basic logging unit. Under the jurisdiction of a *lespromkhoz* are a number of *lesopunkts* or 'independent' logging areas. Usually there are between two and four such logging areas attached to each *lespromkhoz*. Each will consist of an area of forest which it logs, a series of roads, tracks, possibly railways and cableways, depots with the tractors, saws and other equipment required for operations, a repair and maintenance department, a number of upper landings and a *nizhni sklad* or lower landing. The lower landing for each *lesopunkt* acts as a storage depot and sorting yard for the timber logged within its area and as a point of preliminary log breakdown. Usually logs are transported to the lower landing in the form of tree length logs or whole trees with crowns intact. In mountainous areas a short wood logging system is used when the trees would be cross-cut into the log lengths required by the final user on the forest floor. At the lower landing the timber is delimbed (if the whole tree logging system is in use) cross-cut and sorted into size, species and quality categories. In some cases chips will be produced from the branches, tops and low quality wood extracted to the lower landing from the forest. At some of the larger lower landings with an annual throughput of more than, say, 200 thousand cubic metres some further log breakdown is often carried out. For example, railway sleepers, pit props of split pulp wood may be produced. Up to a third of the output of lower landings may be in the form of these 'semi-manufactures' rather than just round wood assortments. According to the figures given earlier, the average annual production for a *lesopunkt* should be between 50 and 150 thousand cubic metres. But, whilst the lower figure is accurate, the upper limit is an underestimate. About 10 per cent of *lesopunkts* manage to produce more than 150 thousand cubic metres and one or two even output about 500 thousand (Zalegaller and Lastochkin, 1973).

A typical *lespromkhoz* structure is shown in Fig. 3.4. Within this particular *lespromkhoz* there is one depot transferring timber from river to railway and four *lesopunkts*, each with its own lower landing. The two types of lower landing that are in existence are also illustrated in the diagram; one is situated on a rail route and the other on the side of a river. The rail-side landing transports timber to the final user by the State-owned rail system whilst the river-side one uses boats or, more usually, floats the timber it delivers. Figures for the size distribution of these two types of lower landing are given in Table 3.2. As may be seen, river-side landings predominate but they are, on average, smaller than rail-side ones. The average throughput of all lower landings in 1972 was 82 thousand cubic metres per annum (Varaksin and Stupnev, 1974).

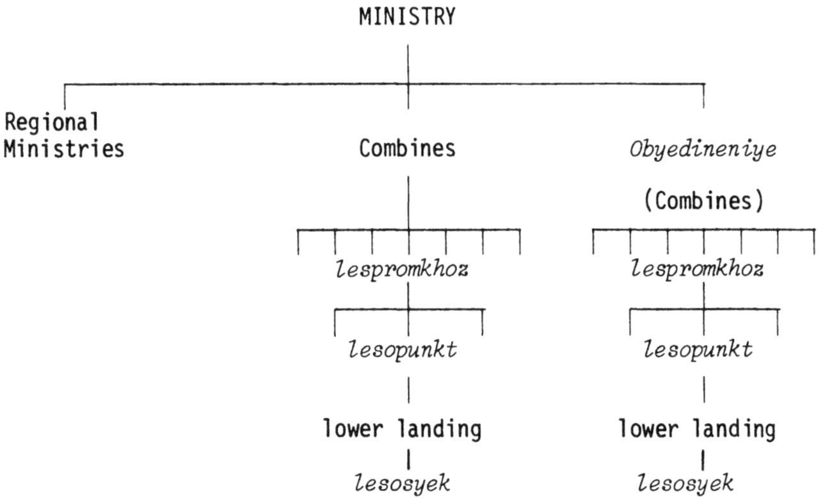

FIG. 3.3 The Management Structure of Logging

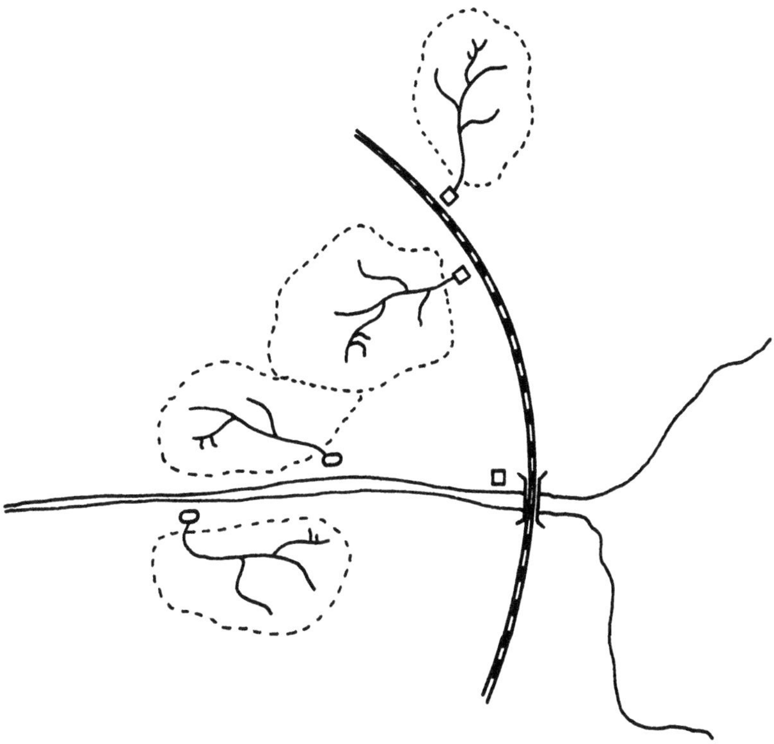

FIG. 3.4 A Typical *Lespromkhoz* Structure

TABLE 3.2 Results of a Sample of Lower Landings

Annual throughput in thousands of cubic metres	All lower landings		Lower landings on rail routes		Lower landings river-side	
	number	%	number	%	number	%
less than 50	1340	49.5	137	19.5	1203	59.2
51 - 100	761	27.6	201	28.2	560	27.3
101 - 150	348	12.6	179	25.0	169	8.4
151 - 200	128	4.6	73	10.3	55	2.8
201 - 250	72	2.6	48	6.7	24	1.3
251 - 300	32	1.1	25	3.5	7	0.3
301 - 350	37	1.4	29	4.1	8	0.4
351 - 400	16	0.6	9	1.2	-	-
more than 400	11	0.4	11	1.5	-	-
TOTALS	2745		712		2033	

The sample refers to the position in 1974. It may be expected that the number of small lower landings has decreased since then.

Source: Kozhin and Styazhkin, 1976

Looking at a *lesopunkt* in more detail, the structure within which logging is carried out will become apparent. In Fig. 3.5 the organisation of a typical rail-side *lesopunkt* situated in a fairly flat area is shown. The lower landing, 1, is situated on the rail-side. The area which the lower landing serves is delineated by line 2 and this area of forest is itself divided into two parts by line 3. One of these two areas is logged during the winter and the other during summer. The area to be logged in the summer will be the drier of the two. In winter in most areas temperatures are so low that the ground becomes hard and skidding and extraction are easier. In spring, summer and autumn this is not the case and working on the forest floor becomes very difficult. The timber felled in each separate logging area (*lesosyek*, 4) is skidded over the forest floor to an upper landing, 5, which is sited on a forest track or *oos*, 6. This is a temporary road with a life of often not more than one year. There are several such tracks and they lead to a temporary forest road or *vyetka*, 7. These usually have a planned life span of between two and ten years. These temporary forest roads in their turn lead to permanent roads (*magistral*, 8) along which the timber is extracted to the lower landing. In some circumstances timber is taken directly to the final user without first being stored, sorted and broken down at a lower landing.

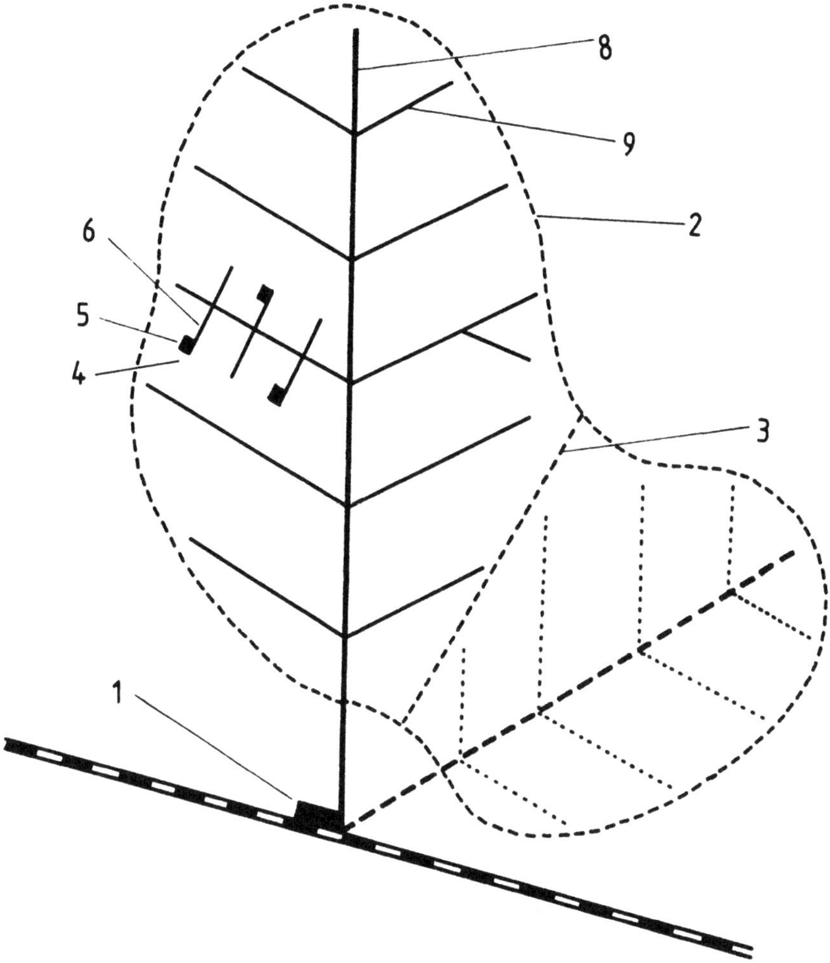

FIG. 3.5 A Typical Rail-side *Lesopunkt*

ASPECTS OF LOGGING IN THE SOVIET UNION

Introduction

A detailed discussion of the actual methods of logging and log breakdown will be presented in the next two chapters with a view to predicting the growth of the productivity of labour in the future and hence estimating the potential of increasing the output of timber in the Soviet economy. To set the scene for this discussion it is necessary now to describe the main means of getting the timber from the stump to the lower landing and beyond.

Engaged in each individual logging area and attached to an upper landing work a number of 'brigades' of four or so men (at the present moment there is a trend to extend the size of these brigades by a process of amalgamation) under the leadership of a brigadier or foreman. These groups represent the lowest level of planning in common use in Soviet forestry and each *lespromkhoz* will break down its annual plan into plans for the various brigades operating in its area. The trees in a brigade's area are usually felled by chain saw, although the investment programme at present being carried out means that an increasing proportion of the timber logged is felled by some form of machine - a feller-buncher for instance. Once felled the tree may be delimbed on the forest floor by means of a chain saw or axe or taken to the upper landing and delimbed there where conditions are likely to be a little easier, even though the upper landing is usually no more than an area of level ground cleared for the operation of delimbing and loading. The timber is skidded from the forest floor to the upper landing by tractor in most cases. Most commonly, the tractor is equipped with a winch and choker log gathering system. In this system a chokerman attaches the timber to the main cable from the winch on the tractor by means of cables with running chokers which grip the boles of the trees as the main cable is hauled in. This particular operation is one of the most arduous in Soviet forestry and is being done away with by the introduction of grapple skidders which gather logs by means of a hydraulically controlled grapple operated by the tractor driver from the relative comfort of his cabin.

Once the timber has been skidded to the upper landing it is loaded onto something for extraction to the lower landing. This loading may not be carried out immediately, the skidded timber may be stockpiled at upper landings for loading at some later date. The loading operation is becoming increasingly mechanised with the introduction of specialised loaders to replace the winch/man power methods in common use now. The most usual means of transport to the lower landing is lorry; narrow-gauge railways were a popular way of extracting timber at one time and are still used in those areas where they have been built, but because of the relatively high capital outlays associated with the construction of such systems, no new railway systems are being built. Lorry extraction follows the series of road types described above, covering a relatively short distance on forest tracks, moving onto temporary forest roads of slightly better quality and finally onto permanent roads of some description.

Once at the lower landing, the timber is unloaded. If the timber has not been delimbed at either the upper landing or on the forest floor it is delimbed here. The timber is then cross-cut into the lengths required by the final users and sorted into various stock-piles to await further transportation. Depending upon the scale of operations at the lower landing, the degree of mechanisation varies. At the smallest lower landings, loading and unloading have a heavy manual input and the cross-cutting and delimbing of the timber will be by hand using chain saws. At the largest lower landings, automatic lines will cut the logs into the required lengths, delimbing will be totally mechanised and, to a large extent, the element of manual labour in loading and unloading will have been removed. The size and degree of mechanisation of a lower landing situated at a river-side is likely to be lower than that of a landing on a rail route, as was pointed out above. Indeed, the smaller lower landings that use water transport to convey timber to the final user are likely to carry out relatively few operations apart from unloading the timber delivered to them and forming rafts of timber for floating down stream. In some the timber will be floated loose and in this case the lower landing may be little more than the bank of a river and a few machines to take the timber from the lorry and place it in the water. During the winter, the work at lower landings such as these might consist of just laying the timber in rafts of some type or loose on the ice that will have formed on the river to await the spring thaw.

In the larger lower landings not only will the degree of mechanisation be greater but so also will be the number of operations carried out. The output from landings that are very large will be more than just timber cross-cut and sorted into the correct species and size assortments. Sawn timber, pallets, boxes and other items may be produced; increasingly chips are being produced at the lower landing from low quality wood that might not otherwise have been used. Often, also, the capability to use the chips produced to manufacture particle board is grafted onto the lower landing making it into more than just a sorting and storing depot. In the newer industrial combines that are being built now or are already in operation, such as those at Bratsk, Ust-Ilim and Sitvitkar, the lower landings have been swallowed up in the integration of the wood-working and wood processing industries found on the one site.

The Effect of the Climate on Logging Operations

As has been hinted at above, the climate affects some of the operations that go to make up logging. One of the main effects of the climate is to make the forest floor far less accessible during the warm periods of the year and to reduce the carrying capacity of the forest tracks and roads during the spring and autumn. To illustrate the extent of this problem, Sucachev, Nesterov and Pagrebnyak have divided the area of *Goslesfond* into four categories. In the first, machines can operate on the forest floor for all of the year except maybe a small period in early spring after the melting of the snows. In the second category the soil characteristics are such as to allow the free movement of machines and workers except during the spring and autumn when temperatures hover around zero. In the third category of forests, tractors quickly cause deep rutting of the temporary forest roads for all of the warm period of the year whilst forests in the fourth category have such poor soil characteristics that the forest floor is impassable during the spring and autumn periods and very boggy during the summer.

Most of the forests of the Soviet Union are to be found in categories three and four. Seven per cent of all forests are in the first category, 36 per cent in the second and 34 and 23 per cent in the third and fourth respectively. Thus, in more than half of the Soviet Union's forests the passage of machines and workers is very difficult or impossible during certain periods of the spring and autumn and in only 7 per cent is the forest floor passable with ease at all times of the year. In some areas the position is worse than this. In *Archangelsklesprom*, *Vologdalesprom* and *Kirovlesprom* the amount of forest in the first category is less than 1 per cent (Vingorov, 1972).

The result of this is that the bulk of logging operations and extraction is concentrated in the more favourable winter season when temperatures are low and bogginess or excess soil moisture is not a problem. This means that, if lower landings are to maintain a steady flow of timber to the final user, large stock-piles have to be created.

The regimes of work of the two basic types of landings shown in Fig. 3.4 (a rail-side and a river-side one) are shown in Figs. 3.6a and 3.6b. In both diagrams line I shows the cumulative delivery of timber to the landing and it may be seen from the slope of the line that deliveries in winter arrive at a faster rate than deliveries in summer. Line II shows the output of timber from the

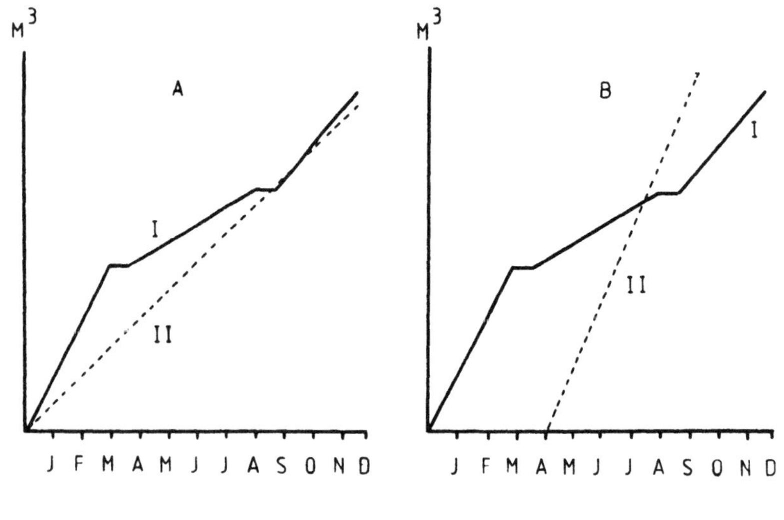

FIG. 3.6 The Work Patterns of Lower Landings

lower landing. In Fig. 3.6a this is a steadily growing output as the transport of timber by railway is relatively unaffected by climatic conditions. However, in Fig. 3.6b the effect of the climate on deliveries is shown. As the timber is transported by river it must be transported at those times of the year when the rivers are not frozen and are sufficiently high to allow the free floating of timber or the passage of timber-carrying boats.

The reserves of timber needed by both types of lower landing to be used during those times of the year when extraction of timber stops or is carried out at a reduced rate can reach up to 7 or 8 per cent of the annual throughput of the landing; this would give an average stock-pile size of about 6,500 cubic metres per landing. If this 7 to 8 per cent figure were followed throughout the industry with production running at about 350 million cubic metres per annum, the total volume stock-piled at any one time would be about 28 million cubic metres.

This problem of the unevenness of the flow of work in Soviet logging is important in many of the decisions being taken about the future development of the industry. A more even flow of production would not only mean that machinery could be used more economically but would also mean that there would be less need for temporary or seasonal labour to swell the work-force during times of favourable working conditions. Also, there would be less need to move workers from their usual jobs to those tasks for which there is a heavy seasonal demand - jobs in which they are likely to be less skilled and less productive. A more permanent work-force, it is argued, is more productive overall and is a necessary pre-requisite for the introduction of the more capital-intensive logging systems.

The reliance on favourable weather conditions for the successful fulfilment of the plans means that the industry, and those that depend upon it for the supply of their raw materials, are susceptible to adverse weather conditions. Recently, in 1977 and 1978, the weather was particularly bad; a long drawn-out autumn followed by a lot of snow in the winter (there were reports of bulldozers being lost in snow drifts) and an early spring and a wet summer made conditions for timber extraction very difficult (Kanevski, 1979). In 1978 the conditions for log floating were such that some wood processing and using concerns had to stop work as they ran out of stocks. In the northern parts of European Russia they experienced, in the winter of 1977 and 1978, colder conditions than the the rest of the country. The coldest temperatures since 1937 were recorded which meant that the spring thaw came about twenty days later than usual. Ice breakers were called in in some areas to open the rivers for the beginning of the timber floating season. To make matters worse, after the thaw arrived weather conditions were such that the rivers fell to levels much lower than was normal. The result of this was to reduce the amount of timber transported and also to increase losses in timber that was floated (*Lesnaya Promishlennost*, 8.6.1978). Eventually, thanks to a slightly longer autumn than usual and the transfer of some timber to lower landings using road and rail distribution, the areas affected registered a shortfall of output for the year of only 400,000 cubic metres (Borisoviets, 1979).

In an attempt to overcome this irregular supply, large stockpiles of timber are formed at lower landings and, in 1977, there were 16 million cubic metres tied up in this way at the lower landings of *Minlesbumprom*. As the Ministry fells about 250 million cubic metres of timber annually, this figure is close to the 8 per cent quoted earlier. However, this level of stock-piling did not allow the lower landings to work rhythmically as Table 3.3 shows. The work at preliminary log breakdown at the lower landing was greatest in the first and fourth quarters of the year and production in the second and third quarters was only carried on at half the level of the first. Considering the changes in these figures over time, it is evident that production is becoming slightly more even through the year.

Bagayev (1979) argues that, on the basis of these figures, many *lespromkhozi* are not forming sufficiently large stock-piles, possibly in the mistaken belief that such a course of action would increase the costs and labour input required in the work at lower landings. He goes on to state:

> ' ... it is necessary once again to dwell on the advantages of creating optimally sized stock-piles with the aim of achieving a rhythmic productive process and an economy of material resources.'

Bagayev presents the figures that are reproduced in the bottom row of Table 3.3 as the amount of work that would have had to have been carried out in the period 1974 to 1976 if no stock-piles had been formed. From this he concludes that, in the first quarters of the years in question, the formation of stock-piles reduced the capacity requirement of the lower landings by 3.5 million cubic metres, this being the amount added to stocks during this period. This saved *Minlesbumprom* about 75 million roubles in investment that would have been required to expand the lower landings' processing capacity by that amount and allowed about 4½ thousand workers to work at logging

TABLE 3.3 The Work-load at the Lower Landings of *Minlesbumprom*

	Average volume of work in millions of cubic metres				
	1st qtr	2nd qtr	3rd qtr	4th qtr	Total
1971 - 1973	83.3	43.4	43.5	55.2	225.4
1974 - 1976	81.0	45.8	44.6	57.2	228.6
Work that would have had to be carried out in 1974 - 1976 if no stock-piles were formed	84.5	44.0	44.1	56.0	228.6

Source: Bagayev, 1979

rather than being needed at the landing. These men, he calculates, would be capable of felling and extracting about 3 million cubic metres of timber in the first quarter and so the formation of stock-piles at lower landings indirectly allows the increase of output from the forests during the favourable winter season.

This last point really highlights the problem that the climate poses for Soviet forestry. The winter season represents a favourable time for logging and transporting timber, a time when the equipment may be used more effectively and so more cheaply. For example, the norms for the load per trip of skidding tractors or lorries are greater in winter than at other times, also average skidding and extraction speeds are greater. So, in winter, there is an incentive to maximise output to economise on costs, and often to take advantage of these conditions crews work two or three shift systems extracting timber. But this very maximisation of output in a short period places strains on other branches of the logging industry by causing an uneven flow of raw material and creating an unbalanced demand for labour.

This problem is illustrated in the technical press where calls for a more even flow of timber by the building of roads capable of use throughout the year are juxtaposed with exhortations for the maximisation of output of logging concerns during the winter months. For example, Kanevski (1979) suggests that *lespromkhozi* should extract about 40 per cent of their annual planned volume of output in the first quarter of the year, and up to 45 per cent for concerns in Siberia. Also, the plan for 1982 called for 41.4 per cent of the output of *Minlesbumprom* to be extracted in the period from January to March (Shalaev, 1982). This extraction should be of timber felled in summer and autumn and stock-piled in the forest as well as of timber felled during the winter months. Stock-piles in the forests of up to 18 million cubic metres are suggested as being needed to allow the extraction lorries to work most effectively during the winter period (Stepanov, 1979). Elsewhere, however, the relatively small output of timber during the summer months is criticised as reducing the regularity of flow of timber, decreasing overall output and increasing costs (Tsibayev, 1980).

The effect that this concentration of work in winter has on costs is shown by a large survey of logging concerns reported by Kozhin and Styazhkin (1976). They show that the reduction of the unevenness of flow of timber to a lower landing leaves costs (in terms of roubles per cubic metre at the lower landing) unaltered. The reason for this apparently paradoxical result is that, whilst a smoothing of the flow of timber from a forest allows work to be carried on without recourse to the extra costs of shift working and allows the transport equipment and other machines to be used more regularly and maintained to better standards, the increase in evenness of output can only be got by increasing extraction during the wetter spring and autumn periods when the use of dirt tracks means that the cost of extraction rises. Also, although the output of the various logging machines is usually higher in winter on an hourly basis, the length of time that they spend working is reduced. Workers are allowed more time to 'warm up' during the working day and the machines themselves might require a 50 to 60 minute warming up period in conditions of severe frost, say -30° to $-40^{o}C$ (Asonov, 1979).

If this is the case, it would seem that there are few economies to be had from allocating the yearly planned level of output to different quarters of the year if the transport infrastructure remains unchanged. Therefore it would seem better to view the calls for a high level of output in the winter months as a short-term expedient and to look for long-term ways of reducing the dependence of the logging industry on the weather conditions, by building roads that can be used all of the year for example.

Roading Problems in Logging Concerns

The extraction distance. One of the reasons for the unevenness of timber extraction is probably that, as time goes on, the extraction distance is increasing. Considering Fig. 3.5, when a lower landing is established, the area immediately adjacent is usually the first area to be logged over. As time goes on the area logged becomes more distant from the lower landing, so increasing the extraction distance. Table 3.4 shows how average extraction distances have increased in the past. Since 1950 the average increase has been about 1.5 kilometres a year and the rate is accelerating. Already the average extraction distance of about 55 kilometres hides the fact that in many cases the extraction distance in individual *lespromkhozi* is more than100 kilometres. It is confidently expected by most experts that the average extraction distance will exceed this figure by the year 2000 (Tatarinov, 1978). This is likely to be a result not only of the process briefly outlined above, but also the effect of closing smaller lower landings and concentrating production in the larger more efficient ones.

As the extraction distance grows, unless road building is of good quality, the likelihood that the extraction route will have to cover areas where soil conditions affect the operation of lorries in

TABLE 3.4 Average Extraction Distances

Year	Average Extraction Distance, km
1935	6.7
1940	9.4
1950	11.9
1961	20.5
1971	31.4
1972	34.1
1973	36.9
1974	41.0
1980	about 55.0

Source: Kozhin and Styazhkin, 1976
Orlov and Skomorokhova, 1981

the wet season increases. So it is probable that, if nothing is done to increase quality, the effect of the weather on logging output will increase. Another factor that would reinforce this conclusion is the fact that, as extraction distances grow, the power of the lorries and size of their loads will become greater. This will put even more demands on the road surfaces to withstand the increased wear-and-tear.

The stock of roads in Soviet forests and their quality. In all, a little less than 100,000 kilometres of roads of various types are in existence in *Minlesbumprom's* enterprises. Roughly, 4 per cent of these roads by distance are *magistral*, 13 per cent are *vyetka* (designed for longer usage than one year) and the remaining roads are of the *oos* type designed for a life of one year or less. This shows the heavy dependence on low quality, short life-span roads in Soviet forestry (Krainov *et al.*, 1972, and Gorbachevski, 1973).

The optimum roading density in the Soviet Union is calculated on the basis of a methodology that looks to minimise the overall costs of logging. As roading density increases, the costs of roading also increase but the costs of skidding and extraction decline as these operations become easier. The sum of these two sets of costs usually leads to a U-shaped cost function with the minimum cost occurring at the optimum roading density - for examples of this methodology in the British context see Johnston *et al.*, (1967). Although a strict definition is not given in Soviet texts, at least one source (Krainov *et al.*, 1972) states that the Soviet Union has an average of one-half of a metre of permanent forest road or *magistral* for every hectare of land being logged or that has been recently logged. This is a very low roading density and is far below the optimum that the methodology described above would suggest. This would mean that, by their own methodology, the forests in the Soviet Union require a large investment programme in road building. The projected introduction of faster wheeled skidders is seen as one way of reducing the need for roading. The slow speed of the crawler tractors that are presently used for skidding means that increasing the skidding distance by reducing roading leads to a significant decline in the productivity of skidding and so to steeply increasing costs.

The usability of roads depends, to a large extent, on the nature of the surface used. Figures for the early 1970s concerning the temporary forest roads showed that 33 per cent were gravel covered, 1.5 per cent were strip roads with strips made from precast concrete slabs, 9.5 per cent were made from wooden strips, 35.5 per cent were earth roads and the remaining 20.5 per cent were winter roads made from snow or ice (Krainov *et al.*, 1972). Of these, only 44 per cent, gravel roads and strip roads of various types, may really be considered to be usable throughout the year. More recent figures show that the percentage of forest roads with a covering that would allow their use throughout the year has increased. Now about 56.5 per cent of roads are permanent in the sense that they are likely to be relatively unaffected by the weather conditions. These roads account for about 55 per cent of all of the timber extracted (Demidov, 1978). Apart from the increase in round-the-year roads, there has been a noticeable decline in the use of wooden strip roads. The reasons for this might be the high expenditure in timber needed to build and maintain these roads - on average 300 cubic metres of timber per kilometre for construction and up to 100 cubic metres per kilometre per annum for maintenance. This would represent, on the basis of the earlier figures, about 60,000 cubic metres per annum for the maintenance of the existing wooden roads with more than 370,000 cubic metres tied up in their construction.

The quality of roads appears to be sharply defined depending upon who is responsible for their construction. Roads are the responsibility of two separate 'departments'; on the one hand *Soyuzlesstroi* is responsible for the construction of forest roads that are paid for from the central investment funds. On the other hand, the individual *lespromkhozi* are responsible for the construction of a large number of forest roads, especially the forest tracks and temporary forest roads, the cost of which is borne by the users of timber products as the cost is passed on in the price of timber. About 90 per cent of the various types of forest roads fall into the latter category. However, it is here that the low quality of road construction is felt, for whilst most authors seem to agree that *Soyuzlesstroi* is competent to build good quality roads, the *lespromkhozi* themselves do not seem to be able to. To quote at length:

> 'Roading is a very difficult piece of engineering. Their surveying, planning, building and maintenance should be the province of specialists. All of this is understood by *Soyuzlesstroi*. Therefore the quality of the roads that they build is in general high. At the same time, many logging enterprises do not even have permanent groups of workers specialising in roading despite the fact that they annually build and use tens of kilometres of roads. As a result, roads built under the jurisdiction of a *lespromkhoz* are distinguished, as a rule, by their low quality. Moreover, in the *lespromkhozi* there is little attention paid to the maintenance of existing roads.'
>
> Vasiliyev and Gryazin, 1979

The authors report that many graduates who have specialised in road building and engineering at college are leaving forestry because they are placed in areas of management other than their specialities because of the lack of special engineering and roading departments in the *lespromkhozi*. In some cases the lack of investment in roading on the part of the *lespromkhozi* is the result of an absence of the capital equipment required. For example, in *Krasnoyarsklesprom* there has been reported a general shortage of bulldozers and graders, two *lesopunkts* having 60 kilometres of roads between them to maintain but no roading equipment. The lack of bulldozers also means that when roads are blocked because of snow drifts it is very difficult to clear them (*Lesnaya Promishlennost*, 17.2.1981).

Investment in roading. The relative costs of roading by either *Soyuzlesstroi* or the *lespromkhozi* themselves are, according to some, good indicators of the differences in quality of the roads built. A kilometre of road built by *Soyuzlesstroi* costs 53,000 roubles whilst a kilometre built by the *lespromkhozi* costs on average only 15,400 roubles. These figures probably also reflect in part the fact that *Soyuzlesstroi* would build roads of a more permanent nature. Other figures for the cost of road building differ from those just quoted. Costs of building permanent forest roads with a gravel surface are given as between 11,000 and 36,000 roubles per kilometre and for temporary roads 10,000 to 18,000 roubles (Krainov *et al.*, 1972).

Roading is also an area where deficiencies in the fulfilment of investment programmes may be seen. The under-fulfilment of the roading investment programme is one of the reasons given by Alekseyev (1982) for the failure of the Ministry to meet the main targets for the 1981 plan. The norms in operation demand that for every one million cubic metres of timber extracted, 8 kilometres of new, permanent roads and 32 kilometres of *vyetka* should be built. In fact, the actual figures are only 23 kilometres overall (Buyer and Korneichuk, 1980). Part of the reason for this lies in the nature of the planning system and this will be discussed in Chapter 7.

Conclusions. As has been mentioned before, the poor quality and small quantity of roads in Soviet forestry leads to a dilemma. That is, whether to invest in permanent roads to allow extraction to continue throughout the year with all the advantages that a smooth supply of timber brings to both the logging industry and the users of timber, or whether to take advantage of the cheapness of temporary roads but to have to pay the price of an output which is not regular. Winter roads are made of either compacted snow or are formed from ice which is produced by pouring water from specially designed lorries onto the road surface (Ilin, 1973). The costs of extraction on such roads is low - almost as low as the cost of extraction on permanent gravel roads and half as much as the cost of using dirt roads (Krainov *et al.*, 1972).

However, winter roads are susceptible to unseasonable weather. Warm spells are likely to cause extraction to cease; in such situations felling usually continues until the winter conditions return with sufficient force to allow extraction to continue. However, when they are usable they offer good technical characteristics for extraction. Speeds can be increased as can loads leading to an increase in the productivity of labour over warmer conditions of up to 25 or 35 per cent. As well as affecting the ease and cost of extraction, winter roads increase the reliability of lorries and reduce the wear-and-tear that they experience.

The future is likely to require an ever-increasing investment in roading. Whilst the skidding distance might increase in the years to come, so reducing the optimum roading density, it is probable that the actual amount of roads will increase. As the size and power of logging lorries increase so the roads built have to be of better quality. Already, single layer gravel roads are only satisfactory if used by the smaller, older types of lorry and more complex and expensive multi-layer roads with greater ground preparation are needed for the newer machines.

Overall, most authors agree that the following points should be put into effect. First, expenditure on roading should increase, especially on roads that can be used throughout the year. Secondly, a better supply of roading materials and capital equipment for road building and maintenance is required. Thirdly, special roading brigades should be set up and used throughout the year, and finally a 'concentration' of planning and executing roading work is called for so that these brigades can be used in more than one *lesopunkt* or *lespromkhoz* to exploit any economies of scale. However, it is probable that the roading requirements of Soviet forestry in the future will be under-fulfilled, as they have been in the past especially as the areas now to come under the axe, around the Angar and Stoney Tunguska rivers in Siberia and the Pechora in the Komi ASSR are all in urgent need of a roading infrastructure.

The Location of Logging Operations

In Chapter 2 it was seen that, in its wider sense, timber transport is very important as wood and wood products have to be moved from the east and north of the country to the regions in the west. This is the result of a mismatch in the locations of the forests and the forest products industries. In the historical past logging was concentrated in the easily accessible populated areas of what is now the Soviet Union and the forest products industries were established there. This was still the case in the more recent past but now the strains of the position are being felt as the forests in the areas where the woodworking industries are situated become exhausted.

Table 3.5 shows the location of logging and, more importantly, the relationship between the amount extracted in a specific area and the allowable cut. In the short-run the allowable cut can be exceeded; the result will be a decline in output in the future when, it is hoped, output in other areas can be expanded. The allowable cut for the whole of the Soviet Union is greater than overall timber removals but in the heavily forested parts of the country the volume extracted is far lower as a proportion of the allowable cut than it is in the lightly forested areas.

In specific parts of the country over-cutting is apparent, especially if the species composition of the timber extracted is considered. The coniferous species are over-cut in many parts of the country, even if in total the timber removals are below the allowable cut. In the European parts of the heavily forested area of the Soviet Union, in the Urals and in several areas of the lightly forested regions there has been over-cutting of the coniferous species, although in these regions the trend is to reduce the amount by which the allowable cut is exceeded.

Lobovikov (1977) views this as a natural long-term process. He notes that the vast majority of logging and wood processing concerns that will be operating in the future have already been built. Whilst investment in new factories will have a marginal effect on output, the existing logging concerns in fact undergo a process of transformation during their life. The initial stages in the life of a logging concern represent a period of heavy logging where the allowable cut may be exceeded until the bulk of logging is transferred to more distant areas. After that there is a reduction in the emphasis on logging and a growth in the importance of more

TABLE 3.5 The Location of Logging Operations (Volumes are in millions of cubic metres)

	Volume Harvested			Percentage of Allowable Cut*					
	1950	1960	1975	1950 1	1950 2	1960 1	1960 2	1975 1	1975 2
Soviet Union	266	369	395	38	53	54	65	57	68
Heavily forested areas	151	253	297	25	45	46	60	52	65
European part	60	58	54	45	76	85	106	91	108
Urals	37	58	54	53	103	88	130	84	115
Western Siberia	12	17	30	14	28	21	34	29	40
Eastern Siberia	25	45	69	15	20	30	29	39	46
The Far East	17	20	33	15	26	24	28	32	70
Lightly forested areas	115	116	98	111	157	106	132	82	106
North west	11	12	14	71	104	67	83	74	96
The Baltic States	8	9	6	173	165	85	118	98	100
White Russia	9	7	6	148	152	130	165	101	104
Central Region	45	41	36	126	170	121	202	87	103
The Povolzhski Region	6	8	9	90	69	106	77	90	92
The Ukraine	15	14	6	204	206	179	192	98	100
The Caucasus	5	6	5	92	200	122	160	90	80
Urals and western Siberia	14	17	14	76	21	79	174	63	93
Soviet Central Asia	2	2	2	51	44	63	61	54	55

* Percentages based on the allowable cut ruling at the time. This varies as it is re-assessed by the forestry authorities. Figures in columns marked (1) refer to total removals as a percentage of the total allowable cut; figures in columns marked (2) refer to the total coniferous removals as a percentage of the coniferous allowable cut.

Source: Glotov, 1977

sylviculturally oriented operations such as thinnings for the improvement of the species composition of the stands and as a source of timber to maintain output until, in the final stages, the forest area attached to a *lespromkhoz* is managed on a sustained yield basis with output coming from thinnings and final felling.

This process is illustrated in Fig. 3.7. The vertical axis, describing the 'degree of development', is not meant to define any precise notion. The present situation can be described as follows. Area I, the area where timber exploitation is in its infancy, is the eastern-most and northern-most parts of Asian Russia. Area II, where logging is well under way but there is little emphasis on sylvicultural operations, contains such regions as *Komilesprom*, *Permlesprom* and *Sverdlesprom*. Area III, where the emphasis is now changing to take a longer term view of the forest resources in the area, would contain *Kirovlesprom*, southern Karelia, *Archangelsklesprom* and parts of the Urals. The other areas, IV and V, would be represented by the Leningrad and Novgorod areas and the southern European parts of the Soviet Union where there are no Group III forests.

This long-term relocation of logging to the east of the country, if it is successful, is going to have several repercussions. First, unless extensive processing capacity is to be opened in the eastern parts of the country, transport distances and costs will rise. Already the transport of timber by rail is causing problems. Railways are an important means of transporting timber over long distances,

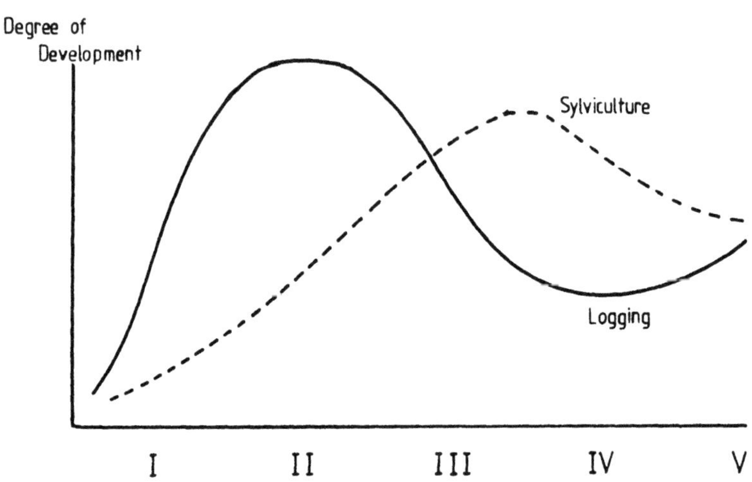

FIG. 3.7 The Evolution of Forest Management

about 160 million cubic metres are shifted annually in this way. But the plans for the distribution of timber by rail are not often fulfilled. In 1977 the plan was only 79 per cent fulfilled and in the first quarter of 1978 the plan was under-fulfilled by 3.2 million cubic metres despite the allocation of an extra 31,000 wagons for timber transport. Part of the reason for this is bad maintenance of the track which reduces speeds and efficiency. Much of the responsibility is the *lespromkhozi's* for they have the duty to maintain much of the track in their area. Another problem is in the loading of the wagons; there is little real mechanisation of this operation and many wagons leave without being loaded to their capacity. Timber has usually to be loaded in the form of individual logs or small bundles of logs because of the shortage of heavy cranes and large panel products are often loaded by gangs of men man-handling the load. The transport of chips, which is likely to become more important in the future, is also causing some concern because the covered wagons which the operation requires are in short supply (Ginsburg, 1978).

As well as the problems of transport, other adverse effects are associated with a relocation of logging. The bulk of the logging operations are moving into areas where the growing stock is poorer in terms of such parameters as average tree size and volume per hectare, and where the climate is more severe. These factors will reduce the productivity of labour in logging operations, increase the wages that have to be paid and so increase costs; in some cases logging towns and villages will have to be built with all of the costs associated with that. Also, it will reduce the appeal of a career in forestry in comparison to other industries so making it more difficult to recruit and hold the work-force.

The *Vakhtovi* Method of Logging

One way of reducing the costs associated with the necessity of logging over more and more remote areas is to employ what in Russian is called the *vakhtovi* method. This is a word that does not easily render itself into normal English but the technique is easily described: instead of settling permanently in an area, a series of small camps are set up with accommodation only for the workers employed in the area and the workers are transported there to work a one, two or three week stint.

The method gained momentum in the early and middle 1970s but first began in the late 1960s (Shchepba, 1980). It was mostly used in the enterprises in Siberia and the Far East such as *Tomlesprom*, *Tumenlesprom*, *Krasnoyarsklesprom* and *Dalesprom* and also in the northern European parts of the country in *Archangelsklesprom*. Its use was not to be the main method of logging but rather was only to be used ...

> '... in cases of necessity, connected with situations when roading is at a very low density, when long transport hauls of the work-force would otherwise be needed, when it is uneconomical to build roads because of poor soil characteristics and in areas where logging operations are situated in isolated areas.'
>
> Yagudin and Gorbachev, 1976

In 1974 there were some 300 logging operations of this type. The technology of working in these types of operations is very similar to that of traditional logging concerns. The areas to be logged are carefully chosen. The parts to be logged in winter will be those where ice and snow roads may be built and where the extraction distance - usually to a river - might be fairly long whilst the summer areas must be sufficiently dry to allow skidding and extraction over relatively shorter distances. In some cases, the method will only be used during the more favourable winter period. The preparation of the area before logging, which is usually carried out by specialist brigades, in this system is carried out by the same workers who will afterwards do the logging; this reduces the number of workers who have to be transported to an area. Logging is carried out in the usual way with an emphasis on the creation of stock-piles on the forest floor for extraction. The separation, by this formation of stock-piles, of the skidding and extraction operations, is seen as very important. Multi-shift working is made use of to maximise the output from the equipment (Gorbov and Zabyalov, 1977 and Reshetnikov and Shugar, 1980).

The men working in these areas live in mobile dormitories which commonly house about eight to twelve people and often six or so of these are situated in the *vakhtovi* village so housing around fifty or more people. Also, in each village, there will be a mobile dining room, reading and rest rooms, mobile electrical generators, workshops, bathrooms, toilets, firefighting equipment and even small shops. As well as these living facilities, all of the equipment needed for logging has to be transported to the area. If the *vakhtovi* village is not too far away from the town where the workers are permanently settled, buses or boats are used to take the workers to the village. This allows all of the workers to be transported in one go and so means that the day of transport might be at least partly used for work; over long distances, helicopters are used.

Although the method is still being used, it is difficult to gauge the full importance of this type of operation either in terms of the costs saved or in terms of the total volume logged. Speaking to some Soviet experts now, a certain disenchantment is apparent. Living in conditions that are bound to be fairly basic is not likely to be very popular and recruiting workers for such operations appears to be proving difficult. Although the wages paid to such workers are higher than usual, it is probable that this sort of work is going to appeal more to the younger worker who has, possibly, less experience and less commitment to the forest industries. As will be seen in Chapter 5 the productivity of such workers is lower than the more permanent workers and the reliability of the machines that they use is less. It seems likely that such a system as the *vakhtovi* method of logging will never provide anything more than a marginal increase in the output of timber from Soviet forests.

The Concentration of Production

One aspect of logging (and for that matter the other branches of the forest products industries such as sawmilling, board manufacture, etc.) that is thought to increase overall efficiency is the concentration of production into a few large-scale enterprises. This has been mentioned before and Table 3.2 showed the sizes of the lower landings in 1974. Since then there has been a tendency to close the small lower landings (usually those which are based on riversides) and so increase the annual throughput from each of the remaining landings.

This concentration has taken place at both the level of the lower landing as well as the level of *lespromkhoz*. In 1950 there were 844 *lespromkhozi* under the control of the main Ministry and these accounted for 132.9 million cubic metres of timber - an annual throughput of about 157 thousand cubic metres. In 1972, however, the number had fallen to 730 whilst the volume of timber harvested had risen to 223.4 million - an annual throughput of 306 thousand cubic metres. Now the number of *lespromkhozi* is in the region of 630, giving an annual throughput of about 355 thousand cubic metres (Moshonkin, 1975).

However, concentration at this level will only yield managerial economies of scale; these might arise in the area of a more rational distribution of output between the various lower landings that constitute the *lespromkhoz*. For example, it might be possible to allocate the production of sleepers to one lower landing rather than having their production spread inefficiently around several. Waste in one lower landing might be of use at another; also there will be some savings in the number of administrative staff needed.

Technical economies of scale, that is to say, economies from being able to process large quantities of timber, will only be achieved if lower landings are enlarged and the appropriate technology installed to enable unit costs to be lowered. The concentration of production also has an influence on labour. The larger the productive unit the larger will be the number of workers required - although not proportionally so or else there would be no benefit in terms of increased productivity. This means that the logging villages have to be larger but, if there are economies in construction, then costs per living unit will be reduced, the facilities can be improved and the attractiveness of life there can be increased.

Important research in this field has been carried out by *Giprolestrans*. The research was based on the costs of a *lespromkhoz* consisting of one highly mechanised lower landing to which all of the timber was extracted. The planning horizon was taken as being forty years. The research suggested that, as far as costs were concerned, the optimum throughput was about 400 thousand cubic metres per annum, whilst the *calculated cost indicator* (this is an index used for investment decision-making and will be discussed in Chapter 8) showed its minimum value at between 450 and 600 thousand cubic metres per annum. This would suggest that about 450 thousand would be considered best from the point of view of Soviet experts. This is very considerably above the present average lower landing size.

The concentration that these figures imply would be desirable would entail the building of more roads as the average extraction distance would increase. This might have some of the adverse effects mentioned earlier. However, the research was interesting in the sense that it showed the existence of diseconomies of scale. Sizes above the optimum result in higher unit costs. This is a comparatively rare observation in the Soviet economy where 'larger' has until recently been synonymous with 'better'; in the past this

this tendency of the Soviet planners to 'think big' has been labelled 'gigantomania' (Gregory and Stuart, 1974). The large complexes, for example at Bratsk, Ust-Ilim and Sitvitkar, are examples of this. Plans for such complexes used to envisage processing capacities in the region of 15 million cubic metres; now half this figure is more the norm. Often the complexes that do exist operate considerably below capacity because of organisational and transport problems. Overall, the degree of concentration required by the figures presented above is unlikely ever to be reached. The construction side of *Minlesbumprom* has a relatively poor record and its performance would have to be considerably better than it has been if the investment implied by the figures for concentration is to be fulfilled.

The Water Transport of Timber

The water transport of timber from lower landing to timber user is still a very important link in the forest industries despite the decline in its use in recent years. In 1950 over half of all of the timber logged by the Ministry was eventually dispatched from the lower landings by river whilst in the 1970s the figure had fallen to around 37 per cent. It seems, however, to have stabilised at this figure (Donski and Savelov, 1973). This represents a large quantity of timber, and in the mid-1970s 2½ thousand rivers and lakes were used for timber floating, their overall length was 140 thousand kilometres and the average distance floated by timber was over 500 kilometres. In certain areas of the country the importance of water transport of timber is greater than the nation-wide average suggests. Around the Pechora river and in the area of northern European Russia in other river basins about 90 per cent of the timber logged is transported by water. In the region of Western Siberia around the Ob river about 94 per cent is moved in this way and in the Amur river basin nearly 70 per cent of the timber felled is floated to the final user (Lebedev and Pominova, 1974).

As this geographical distribution suggests, most of the water transport of timber occurs in the heavily forested parts of the country and in fact about 90 per cent of all timber floating is carried out, at least in its inital stages, by *Minlesbumprom* and only 10 per cent by other authorities (Ovchinikov and Donski, 1978).

Talking to Soviet experts and reading the technical press over the last few years, one gets the impression that timber floating was not looked upon as being a very 'progressive' form of transportation and that its use was being phased out as far as was possible. This was coupled with the point made earlier that many of the smallest lower landings were those which dispatched their timber by means of rivers and so any increase in concentration of production would almost of necessity have led to a contraction in timber floating. However, the reduction of water transport over the last twenty or so years is now seen as a process of rationalisation.

Before discussing the nature of this rationalisation it is necessary to look at timber transport by water in more detail. Timber can be conveyed by water in three ways. First there is loose floating - *molyevoi splav* - where the timber is allowed to float freely under the influence of the current of the river. Secondly timber can be rafted - *plotovoi splav* - where the individual trees are bound together to form rafts which are towed by tugs and, finally, there is actual transport by boat in the conventional sense with timber being loaded onto boats and later off-loaded. Over the past few years there has been a reduction in loose floating and an increase in the other two forms of water transport. Table 3.6 shows this. The

TABLE 3.6 Water Transport of Timber

	Percentage of Total		
	loose floated	rafted	transported in boats
1960	75	23	2
1970	70	25	5
1972	64	30	6
1975	63	31	6
1976	59	35	6
1977	58	35	7
1980*	50	38	12

* This is the planned figure for 1980. In fact the plan was not fulfilled.

Table taken from Ovchinikov and Donski (1978)

reason for this is that loose floating is the least efficient of the three forms of transport. The losses incurred during this operation can be quite high and the damage to the river banks is greater than with other methods of water transport. So its reduction is looked upon as being desirable.

The figures quoted in Table 3.6 represent the percentage figures at the point at which the timber is first transported. The type of water transport measured in percentage terms at the point of arrival is different. During the timber's travel changes in the type of transport are made. Loose floating timber is often sorted and gathered into rafts or loaded onto boats as well. Figure 3.8a shows what happened in 1970 and Fig. 3.8b gives the projected figures for 1990. An increase in rafting and transport by boat is envisaged with a reduction in the relatively expensive and labour-consuming operations of transferring timber from one means of water transport to another.

The advantage of water transport over other means of moving timber, when such alternatives exist, is its relative cheapness. Maintenance of rivers for floating is forty times cheaper than the maintenance of the same length of road and some sixty times cheaper than the maintenance of narrow-gauge railways (Patyakin, 1973). The cost of transporting timber by river is given as one kopeck per cubic metre kilometre compared with 3.2 for the same unit on an

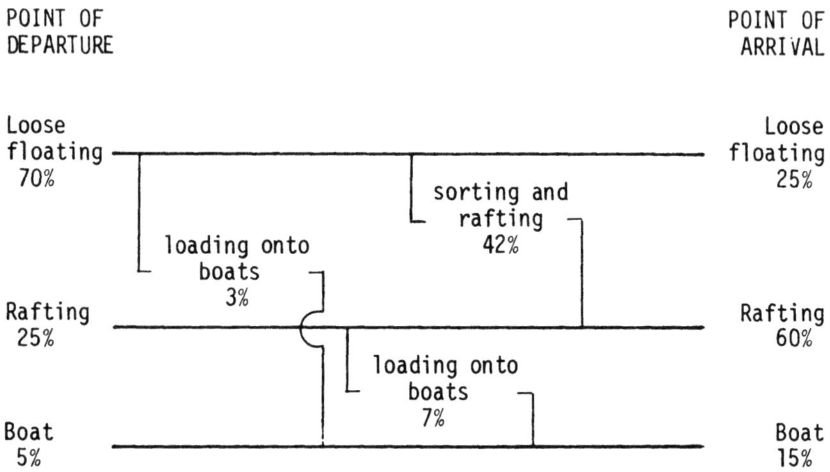

FIG. 3.8a Water Transport of Timber in 1970

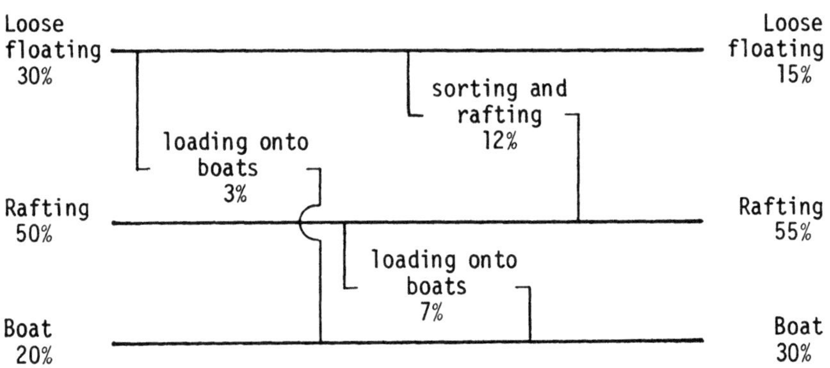

FIG. 3.8b Planned Water Transport of Timber in 1990

Source: Patyakin, 1973

'average' road (Ovchinikov and Donski, 1978). However, the increase in capital intensity implied by the reduction of loose floating shown in Fig. 3.8b will reduce this differential.

Some of the disadvantages of water transport are the relative slowness of the delivery of the timber and the losses incurred in the operation. A little less than 1 per cent of the timber floated is lost (usually sunk) between the points of departure and arrival. Although this is small in percentage terms it represents more than half a million cubic metres of timber per annum. Also, the fact that timber is likely to be lost through sinking means that some species, the lighter ones, are better suited to floating than the more dense ones; this is another reason for the increased use of rafting. A raft of timber can be made up of relatively more dense species in combination with less dense ones, so that the overall density of the raft is such that it has good 'floatability'. This increases the ability of the forest industries to float hardwood species and larches which tend to be denser than pines and spruces. If the denser species are floated loose losses tend to be high. The limited use of rafting until now has meant that much of the timber growing around rivers has been left standing as it is 'immovable'. For example, losses in loose floating of birch over a sixty- to ninety-day period are as high as 20 to 30 per cent, and for ash as much as 50 per cent (Polekhin and Belenov, 1978). Drying the timber before floating is occasionally tried to reduce the density of the wood but this often leads to degradation with a loss of value (Belenov, 1980).

Another problem associated with the transport of timber by water is the seasonality of the work; most of the labour requirement comes in the May to September period when the rivers are ice free - see Fig. 3.6b. Usually the seasonal workers come from the agricultural sector - either from State or collective farms. Clearly this is an undesirable state of affairs for not only are the seasonal workers likely to be less skilled than permanent ones but the demand for labour comes at a time when workers are needed in the agricultural sector. The demand for extra workers in the 'season' has been reduced slightly as more and more timber is 'winter floated' by placing it on the ice in winter, either loose or in rafts. It is then carried down river when the thaw commences; between 25 and 30 million cubic metres of timber are transported annually this way.

However, a possibly more important effect of this dependence of the work on the climate is that the system of transport is never likely to be totally compatible to a Soviet style of planning which allows little room for uncertainty. In the winter of 1978 the climate had an adverse effect on the ability of *lespromkhozi* to fulfil their floating plans; this has been described in an earlier section. In 1979 a similar thing occurred and the plans for timber floating were only 95 per cent fulfilled. In that year some *obyedineniye* lost 5 per cent of their timber during its journey down river and the amount of sunken timber recovered in the period was more than twice the planned figure (Borisoviets, 1980). Also, in 1981, the plan for timber floating was not fulfilled by *Minlesbumprom*. The plan called for 83.2 million cubic metres to be floated; in fact 73.4 million was dispatched from lower landings and only 72.4 million arrived at its destination (Borisoviets, 1982).

In the future it is possible that timber floating will find it difficult to keep pace with the improvements required of it. Already it is noted that the increase in investment in equipment for timber floating has not brought about the desired increases in productivity. Also, as branchwood and low quality wood are likely to be an ever more important source of raw materials for the Soviet timber industries, full tree extraction will become more important. But, not being able to float whole trees in any economic way, the increase in the delivery of trees with crowns intact to river-side lower landings means that capacity will have to be installed to process the tops and branches into chips, say, and transport chips by barge or road away from the lower landing. Some river-side landings are already operating this dual system - water transport of logs and road transport of chips and other 'unfloatable' output.

The move away from loose floating to rafting is to be reinforced by the development of reservoirs and other objects. These affect nearly one-half of all of the water systems used by timber floating enterprises (Chernozhskaya, 1980). Capital investment in such enterprises near reservoirs is 10 to 15 per cent higher than normal to ensure that the rivers are kept 'cleaner' and less timber is lost in the area. Calls are often seen in the technical press for more co-operation between the timber Ministry and the authorities in charge of water resources (Seroken and Filimonov, 1980).

The future of timber floating would seem to require a fairly heavy investment programme. A change from the use of loose floating to rafting and transport in boats requires an increase in the number of tugs and timber carrying boats; as well as that the costs of raft floating are greater than those of loose floating. However, in the past investment in timber floating has been about half that in other forms of transport, despite the fact that it accounts for a little less than 40 per cent of transport from the lower landings.

But the plans for the rest of forestry require a high level of investment - logging technology, living accommodation and villages, timber processing capacity, etc. and so it is likely that the projection for investment implied by the plans for 1990 shown in Fig. 3.8b will be the best that can be attained and even this looks optimistic.

The Baikal-Amur Railway

The Baikal-Amur railway (BAM) is an important new rail line that is designed to improve communications in the Far East. It runs from Ust Kut to Nizhni Angarsk on the northern shores on Lake Baikal to Zeya and then on to Komsomolsk-on-Amur and Sovietskaya Gavan on the coast. This line runs considerably to the north of the existing trans-Siberian railway, away from the Chinese border.

The area that BAM is to run through is extensively forested and it is anticipated that the development of the line will be associated with the development of forestry and timber industries. The forests in the region of BAM cover 141.7 million hectares, of which 71 per cent or 101 million hectares are forest covered. This represents about 13 per cent of the entire forest stock of the Soviet Union. The area that BAM is to run through has been three-quarters covered by surveys on the ground. According to these there is an exploitable volume of 1,030 million cubic metres of timber. The predominant species is larch, accounting for more than half of the growing stock, spruce accounts for about 20 per cent, 15 per cent is pine and cedar and birch are also common (Kirukhin, 1977).

However, the region is far from ideal for forest operations. It is very hilly or even mountainous in places and so much of the forest area is likely to be designated as Group I or at least be under a restricted cutting regime. Around the area of Lake Baikal 20° slopes constitute over half of the land and in other areas of BAM the figure is between 30 and 40 per cent. In the area to the west of Baikal up to 100 avalanches a year have been recorded; also the climatic and topographical conditions are such that erosion is likely to occur in the event of clear-felling operations. One centimetre of soil cover requires 200 to 300 years to form in such conditions (Vorobiev *et al.*, 1979).

The first part of the area around BAM to be exploited is to be that from Ust Kut to Nizhni Angarsk. In this area four timber processing complexes are planned with a combined throughput of 11 million cubic metres. These will be the Ust Kutski, Taurski, Kirenski and Kazachinski combines. This particular region, as well as being the first to be exploited, is also the area with the best species distribution. Half of the growing stock consists of the more valuable spruces and pines. In the other regions developments are planned to come later with more than six other woodworking combines projected (Dyakonov, 1975). As well as providing opportunities for the internal development of the forest industries in the region, BAM will provide the means for increased exports to the Japanese market for the timber from the Siberian forests.

The significance of BAM for the forest industries is difficult to assess. The difficulty of the area and the past performance of Soviet forestry in the field of investment (see Chapter 5) leads one to believe that output is likely to be less than the figures projected. Also, if the evidence to be presented in Chapter 6 is correct, and the forest industries are being constrained by a labour shortage caused in part by the location of the forest industries in unpopular areas of the country, it is unlikely that BAM will open up forestry in the short run. Already difficulties in its construction have been noted because of labour problems and difficulties in recruitment and maybe BAM will only succeed in causing labour to be attracted to work in the area it opens up at the expense of other logging concerns.

Sylvicultural Operations

Minlesbumprom. The operations of *Minlesbumprom* are not confined to logging and preliminary log breakdown at the lower landing. Part of their operations is concerned with sylviculture - that is operations designed to ensure the regeneration of the forest and the growth of the desired species. First, the actual logging operations are often modified to ensure the greater survival of any understorey or advance regeneration that might be present. Skidding is carried out with the tops of the trees forward instead of the stump end first; this increases the survival of the undergrowth from 55 per cent at a distance of 15 metres from the skidding trail to 90 per cent (Gatskevich, 1971). This method of skidding is likely to increase in the future for another reason; it tends to reduce the amount of dirt and stones that the crowns of the trees gather during skidding. So if the full tree system is to be used and the branches are to be chipped this method has obvious advantages.

But more importantly, logging concerns have the responsibility of clearing the area logged over of brash and logging waste. The reasons for this are to increase the likelihood of regeneration by the most desired species on a site after logging and also to decrease the fire risk. If a tree length logging system is in use and delimbing has been carried out on the forest floor there is clearly more attached to this operation than if the full tree system is in use. Usually the brash is gathered into piles on the forest floor and sometimes, after clear-felling, the brash is burnt. It is conidered that the clearance of brash facilitates the regeneration of the area, especially if artificial methods of regeneration are to be used. If natural regeneration is to be used it is usually suggested that the brash should be gathered into small piles and left.

This gathering of brash seems to be still continuing despite the discussion at the beginning of the 1970s that it was both time-consuming and inefficient. Now, the degree of mechanised clearing has increased but in 1969 it was pointed out that about 15 per cent of all of the working time of logging brigades was employed in this operation and that it was adding 88.4 million roubles to the annual cost of timber logged by the Ministry (Urkin, 1969). Also it was suggested that brash clearance was, in fact, counter-productive. For example, it was found that the regeneration of broadleaved species at the expense of coniferous species was encouraged by the clearance of brash. In some experiments it was found that the percentage of coniferous species in the regeneration was 63.5 per cent in areas that had not been cleared of brash but only 15.1 in areas that had been cleared in the usual way. At the same time the total amount of regeneration was unaffected by the operation - it amounted to 27,300 seedlings per hectare on the uncleared plots and 29,000 on the cleared ones. It was also concluded by many specialists that the fire hazard on cleared and uncleared sites was not significantly different (Vingorov, 1969).

However, whilst the sylvicultural reason for brash clearance may not stand up to close scrutiny, now a different reason for the operation is emerging. With the demand for timber increasing there is a need to utilise all of the timber growing on a particular site, including what might once have been considered waste.

Gosleskhoz. The main organ that deals with sylvicultural operations in the Soviet Union is the State Committee for Forestry of the Council of Ministers - *Gosleskhoz*. Under its ultimate control comes 94.6 per cent of all of the Soviet Union's forests - most of the remaining are under the control of various agricultural bodies. It controls about 2,500 organisations concerned with various sylvicultural operations and fourteen or so scientific research institutions.

The links between *Goleskhoz* and the various enterprises that are under its control are similar to those that are evident in *Minlesbumprom*. However the terminology is slightly different.

Instead of *lespromkhozi*, *leskhozi* are formed which have a sylvicultural role. Sometimes *lespromkhozi*, or *loskhozagi* in the Ukraine, are formed when the actual logging of the forests under the enterprises' control is an important aspect of their operations (Yanushenko, 1977).

Gosleskhoz is responsible for a number of operations. First it is invested with the task of the overall management of the forests of the Soviet Union for the national benefit. Secondly, it has to oversee the regeneration and improvement of the forests, this is its main role. Thirdly, it undertakes some logging activities itself, harvesting between 70 and 80 million cubic metres of timber each year. Fourthly, *Gosleskhoz* has jurisdiction over the establishment of forests managed for the benefit of agricultural concerns.

In its first role, *Gosleskhoz* has a responsibility to survey the forest resource and assemble data concerning its constitution in order to be able to manage it rationally. The data collected are used to determine the allowable cut in the Soviet Union's forests. The determination of the allowable cut comes from the synthesis of the data concerning the growing stock and an idea of the optimum rotation length of the various species present. This latter piece of information will be a matter for discussion by the research institutions under *Gosleskhoz* and the ideas concerning this concept are examined in Chapter 8. Part of the area of land consistent with the allowable cut is then transferred to the control of *Minlesbumprom* who are responsible for the felling of the timber in the area and the elementary sylvicultural operations described above. Usually this area of land will be about 2 million hectares each year.

As well as this aspect of forest management, *Gosleskhoz* also has responsibility for the protection of forests from fire and attack by pathogens.

The second role of *Gosleskhoz* is probably the most important in the sense that it absorbs most of the resources that it has available. For the regeneration of the stands of timber after logging, or for the establishment of totally new planting, *Gosleskhoz* has to collect seeds from desirable tree species, prepare plantations and grow seedlings, prepare the land to receive seedlings or seeds if direct sowing is to be used, tend the young plantation as well as to tend areas in which natural regeneration is to occur. A general idea of the extent of these operations is shown in Table 3.7. In the tenth five-year plan, 1976 to 1980, the area of forest that it was hoped would regenerate naturally was 5.5 million hectares and 5.2 million were either planted or seeded (Vorobiev, 1981).

It has already been noted that in areas of natural regeneration, broadleaved species are often the dominant group of species, regenerating at the expense of coniferous ones. This would suggest that the work of *Gosleskhoz* is insufficient and, indeed, in this sphere it does little more than oversee the operations of the logging concerns of *Minlesbumprom*. In the artificial regeneration of forests more has to be done. As can be seen from Table 3.7, about half of the area of forest logged each year has some form of artificial regeneration. This work is unevenly distributed around the Soviet Union; in the European parts of the country more forests are planted than the area of forests that are felled, so increasing the total forest area whilst in Siberia and the Far East artificial regeneration

TABLE 3.7 The Operations of *Gosleskhoz*

		1940	1950	1960	1970	1975
Natural regeneration	1,000 ha	23	343	664	1024	1084
Regeneration by planting or sowing	1,000 ha	232	557	816	1261	1250
Intermediate felling	1,000ha	1096	887	1075	2095	2333
Drainage	1,000 ha	9	25	73	224	320

Source: Anoshin *et al.*, 1977

may not be carried out at all.

The species chosen for regeneration are usually taken from the pines and spruces with oak used in the south, especially in Moldavia. Monoculture is hardly ever used and mixtures, often spruce, pine and birch, are used.

The choice between sowing seeds directly, planting seedlings or using natural regeneration will depend largely upon local conditions. Some evidence suggests that there is very little to choose economically between the methods. If natural regeneration is to be allowed, a system of felling will be used that allows either seed trees to remain standing on the site in question, or else the areas clear-felled are spaced such that they are always surrounded by stands of timber that are left standing long enough to allow the area to seed over. The cost of this, as far as logging expenditures are concerned, is higher than the costs when simple clear-felling is used but the costs of regeneration are obviously less than when artificial methods are employed. The costs of regeneration are more-or-less reduced to ensuring that the species mix on the regenerated area is not too undesirable. The overall balance of costs of felling and regeneration is such that no one system of regeneration has the upper hand in all circumstances (Rumyantsev and Konyeva, 1972). In passing, it is interesting to note that the piece of work just referred to used a discounted cash flow method over the whole life of the rotation of a plantation. This is not a particularly common methodology - see Chapter 8.

When techniques of planting seedlings are used, the methods are similar to those used in the West. Seedlings whose root balls are contained in some medium (peat brickettes, paper pots, etc.) have been developed (Girgidov and Tostopyatenko, 1977) and planting machines that allow the mechanised planting of such seedlings are in use (Ogievski, 1974 and Zima and Matugin, 1976). The mechanisation of this operation, whilst increasing the productivity of labour, has meant that the clearance of logging waste and debris after felling has become more important; this is because, whilst the sowing of seedlings was possible through logging waste when planting was carried out by hand, the machines used tend to find it very difficult to plant through brash.

Very often planting or sowing is preceded by some sort of soil preparation such as ploughing or the removal of stumps of the felled trees. Usually podzols are ploughed to break up any iron pan and to improve drainage and release the nutrients in the lower horizons. In the areas with better soils, ploughing is carried out to control weed species. Boggy regions are not often ploughed it appears, and as a result, the species planted do not do as well as they might. Use is made of various strip and block patterns in the formation of plantations. The species used are often, for example, planted in alternating rows.

The third operation that *Gosleskhoz* is responsible for, harvesting timber, resulted in 390 million cubic metres of timber being harvested in the years of the tenth five-year plan. This is 78 million per annum; of this, 200.6 million were the result of thinning operations and the rest the result of final fellings (Vorobiev, 1981). Much of this timber will, of course, be of a smaller size than the timber felled by *Minlesbumprom* because it will be younger. This will make it less valuable but it could provide an important source of raw material for the production of chips for use in paper-making and the wood chemical industries.

The eleventh five-year plan seems similar for *Gosleskhoz* as it is for *Minlesbumprom*. The amount of work that they will be required to do is about the same as it was between 1976 and 1980. In the period 1981 to 1985 it is expected that thinning operations will yield 200 million cubic metres of timber, the same as during the previous five years. The amount of work involving sowing or planting will remain about the same, at about 5 to 5.5 million hectares. In more general terms, an aim of *Gosleskhoz* is to produce plantations in the European Urals to supply the pulp and paper industry there. At the same time an investment programme in accommodation and other 'social objects' will have to be undertaken in order to attract and hold workers in this branch of industry. In the last five-year plan, for example, 1.2 million square metres of living space was constructed, mostly in the form of small blocks of flats although schools, hospitals and nursery schools were also built. A substantial part of this was financed by the retained earnings of the *leskhozi* - see Chapter 7.

4
The History of Mechanization in Logging

INTRODUCTION

In the next three chapters, as well as describing aspects of the Soviet forest industries that are of interest, a prediction of the possible output of timber from the Soviet Union's forests will be presented. The prediction will be based on ideas of the future productivity of labour and the size of the work-force in forestry. In this and the next chapter the possible increases in the productivity of labour that the present programme of re-equipment in the logging industry allows will be analysed. First, historical changes will be reviewed.

Statistics are available that show the decline in the number of man-days needed to produce 1,000 cubic metres of timber. In 1950 1,450 man-days were needed to fell, skid, delimb and transport the timber to the lower landing and prepare it for delivery by cross-cutting and sorting into the required assortments as well as to do all of the preparatory and maintenance work needed to keep logging, transportation and conversion operations going. In 1975 this figure had fallen to only 470 (Baranovski *et al.*, 1976). This represents an annual increase in the productivity of labour of 4.6 per cent. The usual Soviet breakdown of these figures between basic operations - felling, skidding, etc., and preparatory operations - road building, maintenance, etc., is shown in Table 4.1

TABLE 4.1 The Number of Man-days Required to Produce 1,000 Cubic Metres of Timber Ready for Delivery from a Lower Landing

Year	Total man-days	Basic Operations		Preparatory Operations	
		man-days	% of total	man-days	% of total
1950	1,450	910	62	540	32
1975	470	251	53	219	47

Source: Baranovski *et al.*, 1976

On the basis of these figures, the productivity of labour in Soviet forestry would seem to be lower than it is in the West. It is very difficult to compare performance in logging between countries. The nature of operations that constitute 'logging' differs from place to place; in the Soviet Union, for example, brash disposal is obligatory in most situations and so is included in 'logging'. Also in the West cross-cutting and delimbing are usually carried out on the forest floor whilst in the Soviet Union they are often performed at the lower landing.

Some figures for German forest enterprises in the south of that country suggest a requirement of between 1 and 1.2 man-hours to fell, delimb, debark, cross-cut and skid 1 cubic metre of timber. Average tree sizes for the area in question were 0.7 to 0.8 cubic metres - high by Soviet standards (Loffler, 1978). Converting these figures to make them compatible with those in Table 4.1, this represents between 143 and 171 man-days per 1,000 cubic metres. These figures relate to roughly the same operations as those included in 'basic' operations in Table 4.1 for which, in 1975, 251 man-days were needed. According to these figures, productivity in the West is between 1.5 and 1.75 times higher. However, it should be borne in mind that productivity in Scandinavia and North America is generally higher than that in the central European figures quoted here and that the operations such as cross-cutting and delimbing are performed under more favourable circumstances in the Soviet Union than they are in most situations in the West. So, considering these factors, productivity in the West is higher than that in the Soviet Union, maybe by a factor of two or three.

To move back to Table 4.1, the figures show that the larger part of the increase in labour productivity in the Soviet Union has come from increases in productivity in the basic operations. In the twenty-five-year period from 1950 to 1975 the work input represented by basic operations declined from 62 per cent to 53 per cent of the total. Thus, the implied annual increase in the productivity of labour engaged in basic operations was higher than the overall average of 4.6 per cent - in fact it was 5.3 per cent.

A more detailed breakdown of these figures is available from *Archangelsklesprom* for the years 1963 and 1970 (Krainov *et al.*, 1972 and Siritov *et al.*, 1972); these are shown in Table 4.2. The productivity of labour in the Archangelsk area is usually about 7 per cent lower than the *Minlesbumprom* average; this is because of the more arduous conditions in northern European Russia and the relatively low average tree sizes there. What is interesting from these figures is that the percentage of the total man-days under the heading of basic operations has remained more-or-less constant for the period from 1963 to 1970, and referring to the data in Table 4.1, from 1970 to 1975. Therefore, the decrease in the relative share of felling, skidding, delimbing, etc., in the production of timber occurred almost entirely during the years from 1950 to 1963 and from that date the implication is that increases in labour productivity have been equally shared by all forms of operations in the logging process.

Breaking the figures into two periods, from 1950 to 1963 and 1963 to 1975, shows that the productivity of labour in basic operations rose by 8.1 per cent per annum in the initial period, but

TABLE 4.2 Number of Man-days Required to Produce 1,000 Cubic Metres of Timber Ready for Delivery from a Lower Landing in the *Archangelsklesprom* Area

Operation	1963		1970	
	Man-days	%	Man-days	%
Basic operations	353.1	53.7	305.0	55.6
felling/skidding	185.6	28.2	158.8	28.9
extraction	39.8	6.1	28.9	5.3
work at the lower landing	127.7	19.4	117.3	21.4
Preparatory and Auxiliary Operations	304.4	46.3	243.3	44.4
preparation of the felling area	11.3	1.7		
road maintenance	55.4	8.4		
building of logging roads	28.1	4.3		
maintenance and repair of equipment	137.0	20.9		
transport of workers	12.1	1.8		
other	60.5	9.2		
TOTAL	657.5		548.3	

Source: Siritov *et al.*, 1972

only by 2.3 per cent in the second. This calculation takes account of the 7 per cent lower productivity in *Archangelskprom* compared to the country as a whole. This decline in the growth rate of labour productivity is causing serious concern in the Soviet Union and is one of the reasons why Soviet authorities have become disenchanted with the performance of the forest industries in recent years. Indeed, during the 1960s and into the 1970s the performance of the logging industry and, to a lesser extent, the whole of the wood-based industries, fell behind industry in general. The figures in Table 4.3 highlight this shortfall; also it explains why predictions made in the 1950s that the average output per man engaged in logging should soon rise to 1,000 cubic metres per annum have not been met (Voronitsin, 1957). Even the tenth five-year plan's insistence on an increase in the productivity of labour of between 22 and 25 per cent over the five-year period, an annual rate of growth of 4.1 to 4.5 per cent, looked optimistic. In fact this plan was not achieved and the productivity of labour grew very little, if at all, during that period (Solomonov, 1981). Now, the eleventh five-year plan calls for an increase in labour productivity of only 16 to 18 per cent for the five-year period.

In the next chapter, recent developments in logging machinery and future developments in logging techniques will be discussed and the possibility of repeating the growth of productivity experienced

TABLE 4.3 The Relative Performance of Soviet Logging

Industry	Index of Labour Productivity					
	1960	1965	1968	1970	1972	1975
All-industry average	100	125	148	166	186	222
Forest, woodworking and cellulose industry	100	121	141	157	175	na
Logging	100	112	120	125	130	149

Sources: Moshonkin, 1975 and *Pravda* 14.12.75

in the 1950s will be analysed. Before doing this, however, it is both interesting and instructive to look at the development of the typical logging systems that dominated Soviet forestry in the 1970s and which are only now beginning to be replaced. That is chain saw felling, choker skidding by either tractor or cable system and then extraction by lorry. From this the reasons for the actual pattern of productivity growth may be found and, once future trends are analysed, predictions about future performance can be made.

BASIC OPERATIONS

Some General Considerations

The nature of logging operations (felling, skidding and delimbing) have undergone a profound change, characterised by an increase in mechanisation and an increasing capital intensity. In Tsarist Russia, logging was one of the most backward of industrial sectors, relying on manual labour and concentrating operations in winter enabling peasant labour to be used when farm work was not available. What little mechanisation there was seems to have been confined to the use of narrow-gauge railways for timber transportation and this only in the area of the Urals (Ilin, 1973). The mechanisation process began slowly during the post-revolutionary period, and, although research in this area had been carried on before the war, it was not until after the Second World War that significant efforts were made to equip the logging industry with modern technology.

Figures are available to describe this general trend in post-war mechanisation. For example, the graphs shown in Fig. 4.1 are presented by Varaksin and Stupnev (1974) to describe the technical development of logging in the post-war years. They use 'horse-power per worker' as a measure of capital intensity or the level of mechanisation. As may be seen from the graph, this measure almost doubled between 1948 and 1954, rising from 3.5 to 6. After this period the rate of increase in the measure diminished and, by 1964, there was on average only 7.5 horse-power per worker engaged in logging. The slow growth rate continued until 1970 after which

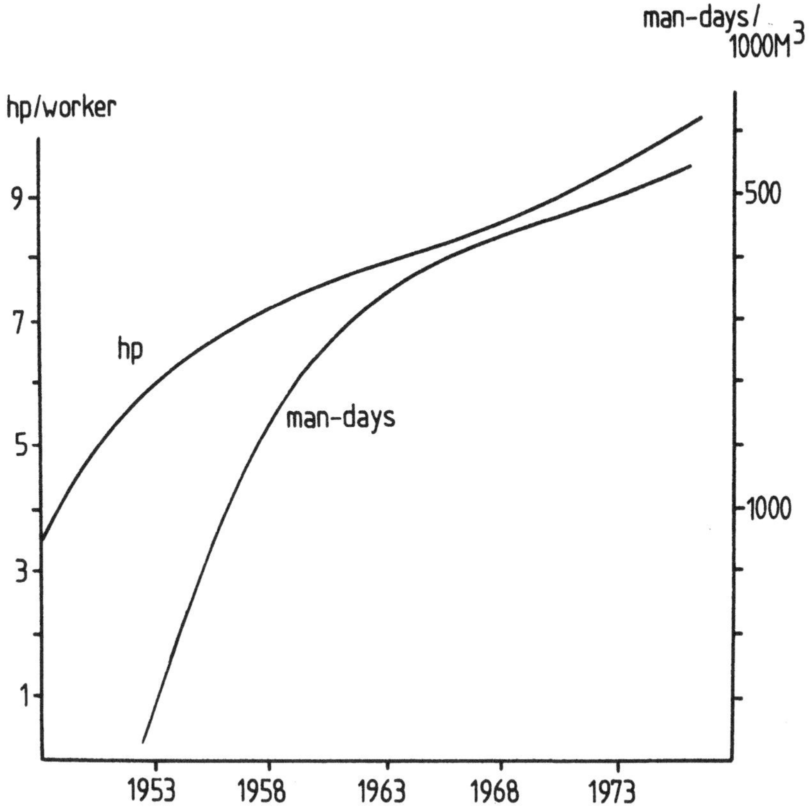

FIG. 4.1 Productivity and Capital Intensity in Logging

there was a more rapid increase until, by 1972, the measure registered a level of 9.4. These figures represent an annual growth rate in 'horse-power per worker' of 9.3 per cent between 1948 and 1954, 2.2 per cent from 1954 to 1970 and 5 per cent over the two years from 1970 to 1972.

Of course, the measure 'horse-power per worker' is a very crude surrogate for capital intensity. Extra horse-power in the form of, say, a substitute of chain saws for bow saws is likely to have more effect on logging performance than the same increase in the form of, say, more powerful logging lorries. So what the measure hides is likely to be at least as important as what it exposes. But, despite the crudity of the measure it is important to analyse this graph for it leads to important conclusions. If the figures for labour productivity are remembered, and they are reproduced in Fig. 4.1, the period of rapid growth of productivity, 8.1 per cent per annum in basic operations, occurred from 1950 to 1963. Other figures suggest that in the period from 1948 to 1954 the performance of the logging industry in terms of productivity was only modest - about 3 per cent per annum (Varaksin and Stupnev, 1974). So most of the increase in productivity occurred in the late 1950s and on to 1963 - after the rapid increase in capital intensity as displayed by this measure. This leads some authors in the Soviet Union to conclude that the graphs in Fig. 4.1 represent a learning effect. The increase in capital intensity from 1948 to 1954 brought about a lagged response in the productivity of labour, the lag reflecting the length of time before the new machines were assimilated and new organisational methods evolved to exploit the possibilities presented by the new technology.

But before speculating further on this point, it is worthwhile to examine the developments that occurred in individual logging operations in the 1950s and 1960s. This will provide us with the information to interpret the relationship in Fig. 4.1 correctly.

Felling

In pre-revolutionary times the forest industries were one of the most backward areas of the economy and felling was carried out entirely by means of either two-handed cross-cut saws, bow saws or axes. At the beginning of the first five-year plan (1928/29) it was reported that only seven motor saws were in use for felling in the entire Soviet Union (Anon, 1947). In the early 1930s (the actual date varies from author to author) the first home-produced chain saws were introduced. There were both petrol-powered saws, probably first produced in 1930, and electric saws; none found wide acceptance because of their weight. The petrol-engined saws weighed about 40 kilograms and the electric types were no lighter. They were, of course, designed for two-man operation but, even so, they were soon taken out of production.

After the war the mechanisation of felling was free again to proceed in earnest. One electric chain saw, the VAKOPP, had been produced by *TsNIIME* during the war years and was ready for general use by 1945. It weighed half as much as the earlier saws and found quite wide usage. However, it was still a two-man saw and, in all, required a three-man gang for felling - two working the saw and one to attempt to guide the tree as it was about to fall by pushing it with a large pole. Productivity in sawing did increase with the use of these saws, but the problem of weight was still there. In an article in a journal of 1948 one logger is quoted as saying,

> 'I've worked at felling for more than eighteen years and I've used two-handed cross-cut saws and bow saws. I could manage, then, on average 3 to 4 cubic metres a day, and after work I was damned tired. Today using an electric saw I can manage about 60 cubic metres ... and after work I feel fine. It would be nice if the Ministry could supply us with lighter saws.'
>
> Rogachev and Shirshov, 1948

The work rate of this logger appears positively Stakhanovite as the technical norm for these saws was in the region of only 20 cubic metres per day per saw. Although this doubtless represented an increase in productivity over manual felling methods, it must be remembered that, overall, the increase in labour productivity would be less than that suggested by the output figures alone. The use of electric saws required an additional labour force to operate generating equipment, which was in fact unreliable and prone to malfunction.

However, by 1948, work to make chain saws lighter was well under way and in that year at least eight different models of electric chain saw were available, ranging from 8 to 20 kilograms in weight and, in power, from 1.6 to 3.2 horse-power. In tests most of these models showed some faults, especially a propensity to overheat, but two - the *TsNIIME*-K5 and the EPKh-3 - performed well.

The K5 was a one-man saw with a standard chain bar arrangement, but the EPKh-3 was an electric bow saw, the chain bar being attached at both ends to a bow. This made it more difficult to use than the K5 and provided a constraint to the maximum size of tree it could be used to fell. Thus, the K5 became the mainstay of the Soviet logging industry until the mid-1960s. It weighed 9.5 kilograms, had a 1.6 horse-power electric motor and could be used by one man. It did not overheat and was convenient to use for felling, delimbing and cross-cutting. Production of up to 40 or 50 cubic metres of felled timber per man-day and, in test conditions, up to 100 were attainable (Minonov, 1948 and Reshetov, 1951).

Mechanised felling using such saws as these accounted for 38 per cent of all timber felled in 1950, and by 1956 this figure had reached 89.6. Of this total about nine-tenths were felled using electric saws, the remainder by petrol-engined ones.

The use of electric powered chain saws, instead of the more normal petrol-engined types, needs an explanation. The petrol-engined saws developed in Russia before the war were very heavy and acquired a very bad reputation for quality and reliability (Uvarov, 1950). As well as this, electric chain saws were probably easier to develop and to make lighter and had less of a vibration problem. But, possibly most important of all, the Soviet authorities had, since the 1920s, been following a policy which favoured electricity over petrol as an energy source for the logging industry. Electricity was used to power winches for skidding and loading, for lighting and experimentally to power skidding tractors, as well as for powering chain saws. This policy in forestry was really a reflection of Lenin's judgement that electrification in general was one of the main tasks of the Soviet government. The generators used in forestry were often powered by burners using logging waste or partially seasoned wood in an attempt to make logging self-sufficient in its energy requirements.

However, there were difficulties in the use of electric saws and, in an attempt to overcome these, the process of development of petrol-engined chain saws continued. These problems were basically two-fold. First, felling performance depended not only on the quality of the saw and its operator, but also upon the generator and the workers operating it. It will be seen later that these generators were often unreliable. Secondly, the feller had to spend time moving the heavy power supply cable around. In conditions of heavy undergrowth this could be quite time consuming, and often where undergrowth and under-storey were particularly dense, they had to be removed or thinned out in the preparatory work prior to felling operations - thus reducing the overall productivity of labour.

These problems were discussed at length in the Soviet technical press and, by 1955, a 2.7 horse-power petrol-engined chain saw (the *Druzhba*) was introduced. It weighed only 10.5 kilograms and had a performance that nearly matched that of the K5. However, official policy still favoured electrification of logging and so it was not until the 1960s that this saw began to replace rapidly electric saws. But it was so successful when it did start to replace them that, by 1970, the only use for electric chain saws was for cross-cutting and delimbing at the upper or lower landings.

The latest *Druzhba* model is shown in Fig. 4.2; it can be seen that its shape is different to the normal style of Western saws. It has bicycle handlebar type handles which allow the saw to be used whilst standing, so lowering operator fatigue. Its design does mean, however, that the saw is unwieldy for either delimbing or cross-cutting. But this may not have been too disadvantageous as delimbing and cross-cutting are carried out under the easier conditions of the upper or lower landing. There either conventionally shaped electric chain saws are used, or axes for delimbing, or if *Druzhba*-type saws are used their chain bar is rotated through about 90° to bring it perpendicular with the ground.

The *Druzhba* is still in wide service but now two other types of saw are also used - the MP-5 Ural, a more powerful version of the *Druzhba*, and the Taiga which is a conventionally-shaped chain saw usually used in mountainous areas. Characteristics of these saws are given in Table 4.4, and from these figures it would appear that Soviet chain saws are still heavier than saws of similar power used in the West - or conversely they are under-powered in comparison. For example, a 5.5 horse-power Western chain saw might be expected to have an engine capacity of about 90 cc and weigh only about 10 or 11 kilograms (Morner, 1976). Both the Taiga and Ural compare unfavourably with this.

The unconventional handle arrangement of the *Druzhba* and MP-5 Ural give good vibration insulation, it would appear. In one book, after testing foreign saws, the authors claim that:

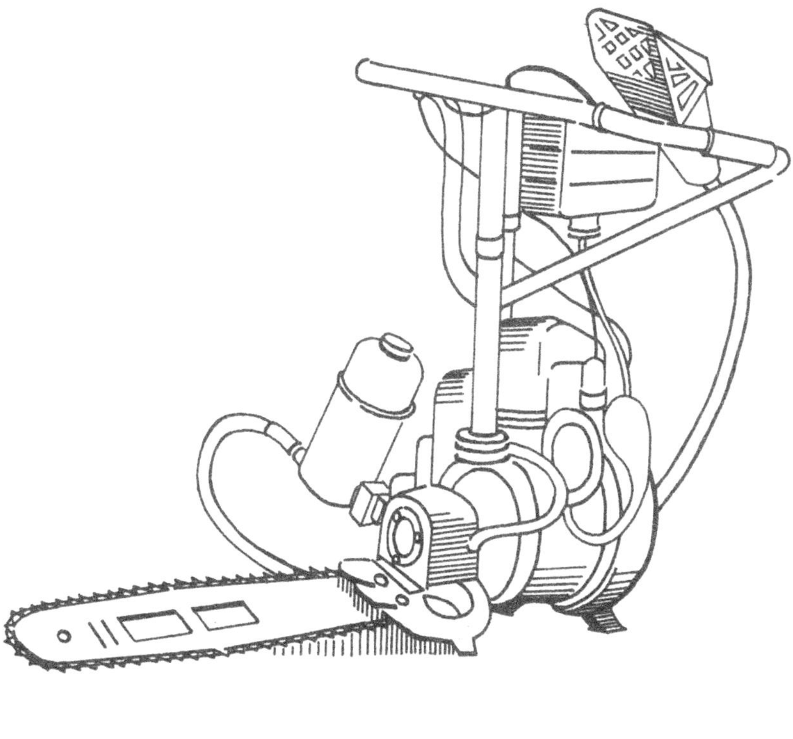

FIG. 4.2 The *Druzhba* Chain Saw and Felling Wedge

TABLE 4.4 Soviet Chain Saws

Characteristic	Name of Saw		
	Druzhba 4	MP-5 Ural	Taiga
Engine power, hp	4	5.5	5.5
Engine capacity, cc	94	109	109
Chain bar length, mm	440	440 or 700	440 or 700
Weight, kg	13.6	12.8	12.0
General noise level, db	111	108	108

Source: Siritov *et al.*, 1972

> 'In spite of the presence of anti-vibration devices ... during the test period ... the foreign saws exhibited greater vibration than the norm existing in the Soviet Union.'
>
> Kochegarov *et al.*, 1970

However, as Western saws are lighter than Soviet ones this probably increases their proneness to vibration problems. It might be that the well-known propensity in the Soviet Union to produce heavier more 'solid' products has here a benefit over the more 'subtle' results of Western engineering.

The introduction of petrol-engined chain saws allowed the productivity of labour to rise as there was no longer any need for operators of electrical generators on the previous scale. However another factor that encouraged the growth of labour productivity was the introduction in the late 1960s of the hydraulic felling wedge. This device, which had been developed by *TsNIIME* as early as 1962, allowed the feller to work alone. These wedges, using the power developed by the chain saw's engine, were inserted in the cut behind the chain bar and allowed the feller to control the fall of the tree. By 1970 about 40 per cent of all felling was carried out with the aid of these wedges.

Using these types of chain saws and wedges, the daily work norm rose to between 60 and 80 cubic metres. But, even in the late 1960s, feller-bunchers and feller-skidders were being experimented with; these were expected to replace eventually manual felling. In the early 1970s, however, they were responsible for only a small percentage of all timber felled and, by 1980, they were expected to account for 30 per cent of all the timber felled by the main Ministry (*Minlesprom*, 1972). These machines will be discussed in the context of the future plans in the next chapter.

Tractor Skidding

Some Soviet authors claim that the crawler tractor was a Russian invention. In 1843, a Captain D. Zagpyazhski patented a a type of crawler track and between 1876 and 1880 F.A. Blinov built and tested 'the world's first automobile running on crawler tracks' - a steam powered tractor (Artamonov and Michaelovski, 1975). However, even if the Russians were the actual inventors of the crawler tractor, they did not apply the technology to the logging industry for a long time. Until the end of the 1920s skidding was totally unmechanised; horses or human muscle power were used to get the timber from stump to roadside. Even in 1926 only 126 tractors were in use in the Soviet Union's forests and these were probably mainly used for hauling sleds for extraction to the lower landing (Anon, 1947). By 1935 only 9 per cent of timber was skidded by tractor. The tractors used were agricultural types powered either by ordinary diesel engines or engines running on gas made from wood in a gas generator mounted on the tractor. As mentioned earlier, planners wished to make logging self-sufficient from the point of view of energy, using logging waste and low quality wood to act as a power source for the logging equipment. One way to achieve this was to use such wood to generate electricity to power electric saws, but another was to use it directly in the production of gas for modified internal combustion engines. In 1937 there were about three and a half thousand gas powered tractors and lorries in use (Gatskevich, 1977).

However, it was not until after the war that interesting developments in tractor skidding were made. The use of agricultural tractors for skidding had obvious disadvantages, one of which was that considerable resistance was generated by the timber during skidding as the load was not suspended at all but was only dragged along the ground. No sulky or arch was used with these tractors for it was found that their use was disadvantageous under Russian conditions as the manoeuverability of the tractor was reduced and the weight of the arch itself added to the resistance experienced by the tractor. Soviet experts concluded that arch and sulky methods were only of use under conditions of 'large average tree sizes with a small number of stems per hectare' (Roos, 1948).

In 1944 it was decided that a special logging tractor should be designed and built. The engineering departments of the Kirov Forestry Academy and *TsNIIME* worked on theoretical designs and, by 1947, a prototype was demonstrated. This was the KT-12 30 horsepower logging tractor; Fig. 4.3 shows an illustration of this tractor.

The KT-12 was powered by gas obtained from a gas producer mounted on the tractor itself. Originally the gas producer used small blocks of wood dried to a moisture content of between 20 and 25 per cent; however, it was found that the preparation of the fuel for use in the gas producer was too time-consuming so a new gas generator was developed which could use green wood with a moisture content of up to 100 per cent on a dry weight basis in sizes up to 50 centimetres long and 8 centimetres thick. Refuelling was needed every one and a half hours and the tractor used a little less than one cubic metre of wood every eight-hour shift. Starting from cold took about a quarter of an hour and a warm engine required less than one minute to re-start.

The Kirov tractor factory in Leningrad was given the responsibility of producing this tractor and, by 1948, loggers had already

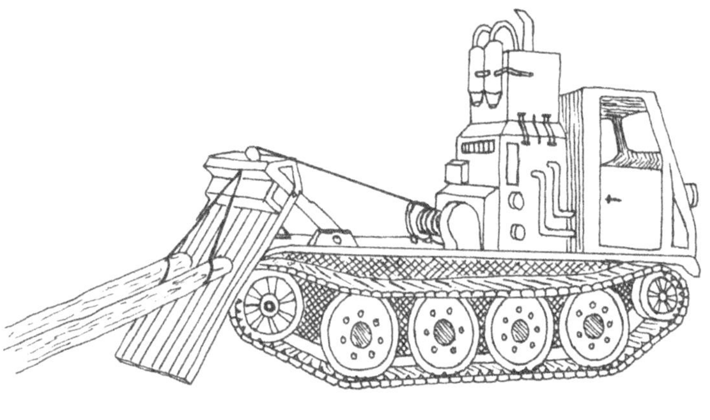

FIG. 4.3 The KT-12 Skidding Tractor

received their first 100 KT-12 tractors. The Russians are quite proud that this was the world's first specially designed and mass-produced logging tractor.

The most interesting feature of the KT-12, which has been maintained in more-or-less the same form on ordinary winch equipped skidders ever since, is the logging plate at the rear of the tractor. In Fig. 4.3 this plate is shown in the down position; the tractor positioned itself on the forest floor and lowered this plate. Cable was played out from the winch located behind the cabin on the main chassis of the tractor and the usual choker system was used to gather the load to the tractor. The main cable ran through a guide at the top of the plate and, as the load neared the tractor, it was lifted clear of the ground with the butts of the trees coming to rest about half way up the plate. By an arm and hinge arrangement the load and the plate were raised onto the rear of the tractor. This method of loading has several important features. First, it provides semi-suspension of the load during skidding; also, the tractor's centre of gravity is moved forward when loaded and its down pressure increased. This increases its traction and pulling power and reduces its tendency to rear whilst skidding. The manoeuverability of this type of tractor is greater than that of a tractor in combination with a logging arch. As well as these advantages, the plate also acted as an anchor during the winching-in process and facilitated the assembly of the load.

In the beginning, as the KT-12 was introduced, there were organisational problems and reports were published that showed a degree of unpreparedness in the *lespromkhozi* for the new tractors. Before the change in the gas generator, fuel caused many problems. Insufficient fuel stores were prepared and additional labour was required to overcome this problem. Down-time measured in days was recorded as the split fuel was left to dry to the required moisture content (Anon, 1949).

However, the tractor was widely accepted and seemed to give better service than the machines, horses or man-power which it replaced. Its good manoeuverability and drive characteristics in the damp conditions of the Russian spring, summer and autumn allowed mechanised skidding throughout the year and so helped to begin to convert logging from a seasonal activity using only temporary labour to a round-the-year operation with a permanent, skilled work-force. Figures for the performance of the KT-12 naturally vary; the average load per trip was usually about 3 to 4 cubic metres, representing between 5 and 10 stems and the winter load was often double that which the tractor could skid in the summer. Output per machine shift varied and some brigades could achieve up to 50 cubic metres but, with an average skidding distance of between 200 and 500 metres, the usual shift production seems to have been a more modest 20 or 30 cubic metres.

At the same time that the KT-12 was being developed, an interesting project was under way to electrify the skidding process. As was mentioned earlier, electrification was viewed as the best way of mechanising many processes in Soviet industry and already in agriculture electric tractors were being used. In 1950 *TsNIIME* developed the ETT-1, similar to the KT-12 except that it was powered by a 38 kilowatt, 52 horse-power electric motor. It received its power by a cable contained on a drum mounted on the tractor which held about 800 metres of cable and tension was maintained automatically. It was played out and taken in through a pulley and arm mounted high on the tractor behind the cabin which ensured that the tractor did not run over its own cable. Tests showed that the tractor was quite versatile and productive. Although it had to return to the upper landing by the same route along which it had left, this did not pose any problems during clear-felling operations. Tests showed it to be about 40 per cent more productive than the KT-12 and the operating costs were about one-third less - requiring only 2 kilowatt-hours per cubic metre transported over an average skidding distance. Many advantages of the electric skidder were noted: it started at once instead of after a quarter of an hour which the KT-12 required; it had no cooling system which caused trouble in Russian winters and required no specially heated garages for its winter storage (Kapitonov, 1952).

However, by 1959 the ETT-1 was still only in experimental form or semi-commercial use and one commentator (Koroleff, 1959) stated that it was still not clear if and when it would be produced for full commercial use. In the event it was not. More conventional tractors were developing more rapidly than the ETT-1 and by the late 1950s the trend in logging was away from electrification. Petrol-engined chain saws were replacing electrical ones for felling and so the need for large generating equipment at the upper landing was being reduced. In that sort of situation the ETT-1 could not compete against more 'independent' diesel-engined skidders.

Returning to the KT-12, problems in its use were being observed. Despite its design there was still a tendency to rear up on skidding and winching-in. Problems of starting from cold remained, fuel preparation was time-consuming and the gas generator had to be carefully serviced or else the tractor lost power. Although thousands of KT-12 skidders had been produced their production was stopped in the mid-1950s; at that time the TDT-40 started production. Built with the same frame as the KT-12, it was powered by a 42 horse-power diesel engine. For starting it had a 10 horse-power petrol engine which itself was started by a conventional electric starter. The skidding equipment on the tractor was the same as that for the KT-12, a single reversible drum winch and logging plate.

Tests on the TDT-40 showed that an improved weight distribution enhanced the manoeuverability of the tractor and prevented rearing up during winching in. Comparative tests showed that the TDT-40 was between 30 and 50 per cent more productive than the KT-12. As a result, the diesel tractor rapidly ousted the wood-burning ones from use and a conversion programme, converting KT-12s to TDT-40s was undertaken and was supposed to have been completed by 1958.

With the KT-12 out of production and the ETT-1 project dropped diesel power was used exclusively in skidders and has been to the present day. There are probably many reasons why diesel engines took over from gas powered ones. The inherent difficulties in the gas generators themselves have already been mentioned; as well as this, the design potential of gas-burning engines was probably less than that of diesel engines, especially if the former were being developed only in the forest industries. Also, there has been a trend away from using low quality wood as a fuel and towards using it as a raw material for conversion in the wood processing industries. Whilst this trend had only just begun in the 1950s it is likely that the planners eventually came around to thinking that, in the long run, diesel engines were more economical in terms of their ability to supply timber to the lower landing instead of consuming it to produce energy. As well as this, of course, by the 1950s the situation in the oil industry had normalised after the effects of the war.

Since the success of the TDT-40 until just recently, when new more radical logging machines have been introduced, the pattern in the design of skidders has been one of increasing engine power whilst keeping the fundamental design unaltered. In Table 4.5 the characteristics of the main winch and choker equipped skidders are listed. Except for the T-100M which is an ordinary industrial crawler tractor which has had a winch system added to adapt it for skidding, all of the tractors in the table have the same sort of logging plate as the KT-12 and use a winch and choker system to gather logs. They differ only in minor design points and in engine power. This is a relatively important feature of the trend of mechanisation in logging which will be examined in Chapter 8; it exposes peculiarities in the investment decision-making process.

In 1972, according to government statistics, the TDT-40 and TDT-55 accounted for 42 per cent of the main Ministry's output of timber whilst the TDT-75 and TT-4 skidded 44 per cent. The former pair of tractors tended to operate in the north-west and central European parts of the Soviet Union, in forests with average stem volumes of up to half a cubic metre, while the latter two machines worked the forests with larger average tree volumes, usually the forests in the Urals and in Siberia and the Far East. The remaining 14 per cent of timber logged by the main Ministry was skidded either by agricultural tractors, in forests with fairly small-sized timber, by the T-100M, in forests with very large timber and by means of winches in the mountainous areas of the Caucasus and Carpathians (*Minlesbumprom*, 1972).

By 1974 the main timber Ministry had over 45,000 tractors employed in skidding, 5,000 of which were in use in mountainous regions where winch systems might otherwise have been used. It is generally considered feasible to use crawler tractors on slopes up to 15° in all conditions, and on slopes between 15° and 25° during dry periods in summer. On slopes steeper than 25° tractor skidding is not usually allowed and it is found that the productivity of tractors on such slopes can be as much as 50 per cent lower than when they are used on level ground.

Most Soviet experts would consider that the tractors now in use in logging are under-powered; research by the Kirov Forestry Academy and other institutions has shown that, as a general rule of thumb, the power of a logging tractor's engine in horse-power should be at least ten times the weight of the tractor expressed in tonnes (Moshonkin, 1975 and Voronitsin *et al.*, 1972). The tractors listed in Table 4.5 do not reach this standard and are also under-powered by Western standards. Some of the new tractors that are now under development meet this norm, and they will be described in the next chapter. But of the tractors in common use, only the T-100M and the TT-4 come close to reaching the required power level, and the

TABLE 4.5 Characteristics of Tractors in Current Use

		Type of Tractor				
		TDT-40	TDT-55	TDT-75	TT-4	T-100M
Length	mm	4500	5850	5505	5950	4255
Width	mm	2014	2300	2370	2500	2460
Height	mm	2430	2245	2700	2750	3040
Weight	kg	6500	9500	11000	12500	11400
Ground pressure	kg/sq cm	0.45	0.41	0.42	0.42	0.49
Maximum load on bunk	kg	2000	2000	4500	6000	-
Engine power	hp	42	62	75	95-130	108
Speed range	kph	2.3-10.3	2.5-10.9	2.3-7.6	2.3-10	2.3-10.1
Drawbar pull range	kg	4440-500	4900-480	6820-1150	9700-1400	9250-1515
Load per skidding trip	cub m	5-7	7-8	12-18	15-20	10-12
Average productivity per shift	cub m	35-40	40-50	70	70	60-80

Source: Anon, 1973; Gatskevich, 1971; Kanevski and Voronitsin, 1974; Perfilov, 1974a and Siritov *et al.*, 1972.

T-100M is being phased out of use as it is not a specialised logging tractor. In 1964 nearly 7,000 T-100M tractors were in use whereas now the figure is nearer 2,000.

Before moving on it is necessary to say something about the fact that crawler tractors have dominated Soviet logging and wheeled tractors have remained virtually unused until now. This is a situation which is the reverse of the position in many countries in the West where wheeled skidders have found widespread use. There is no doubt that military expertise in tank production helped in tractor production in the Soviet Union after the war; indeed, a 'noted Soviet tank constructor, winner of the Stalin prize, D. Ya. Katin' was put in charge of production of the first KT-12s to be produced in Leningrad (Roos, 1948). Also only recently have tyres suitable for the wet soil conditions often encountered in Russian forests become available. Crawler tractors also partly fulfil the need of the logging concerns to comply with the requirements of the forestry service to break down logging waste after felling operations. Finally, Russian forests tend to be characterised by small average tree volumes and high stumps tend to be left when felling in conditions of snow. This results in a large number of small but high stumps left on the forest floor which would, Soviet engineers claim, have caused much more damage and wear to wheeled tractors than to crawler ones. So crawler tractors were the first to be developed for logging and the position has remained unaltered until recently; now wheeled tractors are under development and are being introduced in the hope that their increased speed will enable skidding distances to be increased and investment in road building decreased. This will be discussed in the next chapter.

Delimbing

Delimbing remains one of the least mechanised of all the basic operations in the logging process. By 1971 the level of mechanisation did not exceed 10 per cent. Delimbing by axe or chain saw is still the rule although two types of purpose-built hand-held delimbers are in use, the electric RES-2 and the petrol-engined BS-1. In many cases it is women who perform this operation, 'many thousands' of them being employed at delimbing (Voronitsin, 1975).

The location of delimbing varies; before the war and the widespread introduction of tractor skidding, it was generally performed on the forest floor. But in winter, with a deep snow cover, this was a very laborious process and so now it is usually carried out either at the upper landing or, if there is a demand for branches and tops at the lower landing, delimbing is carried out there. In some areas an intermediate store between the upper and lower landings is formed and delimbing is performed there (Moscvin, 1976) although some experts are now of the opinion that this is not beneficial (Danshchikov and Mikhailov, 1977).

The Electrification of Logging

In the 1950s one of the directions that technical development was expected to take was the almost total electrification of logging operations. One author wrote;

> 'Total electrification produces the most favourable conditions for round-the-clock logging, for an increase in labour productivity and for the further improvement in the living conditions for workers.'
>
> Shahniashvili, 1951

Indeed, this author looked forward to the electrification of narrow-gauge railways used in timber extraction from upper to lower landing as being the completion and logical conclusion after the use of electric saws and tractors or winches.

Electrification had been one of Lenin's ideals, but in post-war Russia with an oil industry not back to normal, electrification offered the added bonus of making the logging industry self-sufficient in terms of much of its power requirements - using logging waste to produce electricity. The phrase being quoted in the press at the time was that the forest 'could and should log itself' (Koroleff, 1959). Figures from 1950 suggest that a *lespromkhoz* using electric saws, winches for skidding and loading at the upper landing and using electricity for cross-cutting, sorting and loading at the lower landing required a little more than 5 kilowatt-hours for every cubic metre of timber ready for delivery from the lower landing (Muzukin, 1951). Of this about 2 kilowatt-hours were required for operations on the forest floor and at the upper landing (Parfenov, 1948).

By the early 1950s many portable generators were in use, and whilst some were diesel or petrol-engined, others were fuelled by gas producers similar to those used on the KT-12. The emphasis on electrification in the Soviet logging industry led to the situation that in the 1950s not only was electricity needed for lighting, but often also for electric saws, skidders, winches and railways. As most of the logging areas were remote from commercial supplies, portable or semi-portable power stations had to be used. Only rarely could State power lines be used despite the fact that these offered the cheapest and most reliable source of power.

However, now, except at lower landings, electrical power is only used for lighting, sometimes for cross-cutting at the upper landing and for some electric winches still in use. Diesel and petrol engines power all the chain saws used for felling, all the tractors and most of the winches in use. Loading is now carried out either by winch system or by specially designed diesel-engined loaders. A lot of the electric tractor units used on the narrow-gauge railways have also been changed for diesel ones.

The reasons for this policy reversal are that, in the final analysis, electrification did not bring about the increase in labour productivity hoped for. Once the weight and reliability problems of petrol-engined saws were solved, they proved more popular than the electrical saws which they rapidly replaced. This was partly because of the effort that was needed to drag cables around with electrical saws. The electric tractor never went into production. However, the actual production of electrical power itself seems to have had inherent difficulties which, even if the electric saws, winches and tractors had proved successful, would probably have led to the abandonment of plans for the total electrification of logging. This problem was mentioned in an article of 1950.

> '... at the present moment electrical generators are working in nearly all the Soviet Union's logging enterprises ... and it is possible to consider them to be firmly established products. However, unfortunately their quality leaves much to be desired.'
>
> Laiko, 1950

Laiko noted that the voltage and frequency of the output of the generators was not stable, the smaller ones were open to the elements and in the winter this meant that repairs took a very long time, their cold starting characteristics were not good, and the large generators were housed in trailers and had inadequate ventilation so making working conditions in the summer very arduous.

As well as these problems, the use of a generator introduced a critical element into logging operations; when it broke down all work had to stop. In winter conditions especially, delays were not uncommon. This meant that either the power generating equipment had to be duplicated (an expensive process) or that the generator had to run with perfect reliability. As well as this problem, there was a large man-power requirement to run these generators - in one *lespromkhoz* thirty to forty workers were engaged in such work (Muzukin, 1951). In some operations 10 per cent of the work-force were employed on portable generators (Koroleff, 1959).

These reasons were probably what eventually caused the change of heart of the Soviet planners and ruled out the use of electrical power. Electricity is still occasionally used for cross-cutting or delimbing at the upper landing if power is close at hand. Now the only real demand for electrical power is to be found at the lower landing where large stationary generators or power from commercial lines may be used.

Loading

Mechanisation of the loading of timber onto lorries, rail wagons or carts at the upper landing did not really begin until after the war. Before this it was a highly labour-intensive and tiring task, especially as more and more of the timber felled came to be transported in the form of tree lengths instead of shorter logs.

With the great increase in the number of tractors and winches used in logging in the 1950s the process of skidding changed from being the movement of one or two stems at a time to the moving of large numbers of stems in bunches formed by the skidders. This produced a similar change in the types of equipment used for loading at the upper landing. In the early 1950s various types of cranes had been used for loading, mounted on rails, lorries, crawler tractors or just free standing. As the tree length system, and to a lesser extent the full tree system, gained in popularity (transferring the operations of cross-cutting to the lower landing) the move was made to simplify the upper landings. Thus, systems evolved that did not use different types of crane or other equipment but rather utilised the winch system on the skidder itself to load the lorries or rail wagons.

Loads of timber were accumulated at the upper landing until sufficient had been skidded there to load on to the lorry or wagon.

When this was achieved, the load was formed into a bundle by wrapping cables with chokers around it and, using two A-frames and the winch on the tractor, the load was placed onto the carrier. In this way, it was easy and inexpensive to achieve the mechanisation of loading and, in areas where tractor skidding was used, this method found almost universal use. If winch systems of skidding were used, cable cranes were more usually employed. By the mid-1960s mechanised loading systems such as these accounted for about 96 per cent of all loading at upper landings and, according to *TsNIIME*, this represented a greater level of mechanisation than that existing at the lower landings (Siritov *et al.*, 1972).

But despite this mechanisation, manual labour was still needed as an input into the loading process to fix the chokers and release them and to guide the load, etc. Usually at least two men were employed on the ground to help with the loading. Although Koroleff (1959) stated that not more than five minutes were usually needed to load the carrier, Soviet authors (see for example, Siritov *et al.*, 1972) suggest that up to 30 per cent of the tractors' time was taken up with this type of loading work. This meant that not only was there a reduction in the amount of timber that could be skidded, but also that the chokermen would be idle for at least 30 per cent of the time. As a result of this, these methods are now being supplanted by more usual types of loaders. Since the 1960s there have been moves to separate the skidding and loading operations and either self-loading lorries are used or loaders mounted on crawler or wheeled chassis. This latter type of loaders will be discussed in the next chapter.

Extraction to the Lower Landing

The extraction of timber from the upper to the lower landing was one of the first areas of logging operations to be mechanised. In the Urals a few narrow-gauge railways with steam engines were operating in Tsarist times to haul the timber for charcoal production for use in the steel factories in the region.

Until the war, however, the rate of mechanisation of timber hauling increased slowly and by 1940 only 32 per cent of timber logged was extracted by mechanised means. In 1930 the first all-year roads were used; they were built by laying rough-hewn cants on consolidated earth tracks. First horses and then tractors provided the motive power. In 1932 the first lorries were used in extraction. Cable-ways formed a part of the overall method of mechanisation but were only used over fairly short distances. Narrow-gauge railway building got away in earnest after 1930 and in the mid-1930s even wide-gauge railways (152.4 cm) were built to transport timber in some areas. In 1940 25.3 million cubic metres were hauled by mechanical means and of this total 10.4 million were hauled by tractor, 9.6 by lorry and 5.3 million by railway and cable systems. The other 67 per cent was hauled by horse.

After the war the process of mechanisation got under way. There was large-scale investment and various experimental extraction systems were tested. Table 4.6 shows the relative development of the various types of extraction over the period 1932 to 1975. By 1966 the extraction process was almost totally mechanised and the dom-

TABLE 4.6 Extraction Methods

	Percentages of the total volume extracted by the methods below										
	1932	1940	1945	1950	1955	1960	1966	1968	1969	1970	1975
Lorry	0.3	12.3	11.2	33.7	40.8	61.4	69.0	70.1	72.4	76.0	83.0
Narrow-gauge railway	0.6	6.6	9.0	16.2	27.3	27.6	24.8	22.7	22.2	18.7	
Tractor	1.9	13.7	6.8	6.8	10.4	-	-	-	-	-	
Cable	0.3	0.2	-	-	-	-	-	-	-	-	
Other	-	-	-	-	-	5.5	5.7	6.8	5.1	5.3	
TOTAL MECHANISED	3.1	32.8	27.0	56.7	78.5	94.5	99.5	99.6	99.7	100	100
Horse	96.9	67.2	73.0	43.3	21.5	5.5	0.5	0.4	0.3	-	-

Source: Krainov *et al.*, 1972 and Baranovski *et al.*, 1976

inance of the use of lorries over narrow-gauge railways had been firmly established. The use of cable-ways was really only suitable over a fairly short distance and so rapidly fell from use with the increase in extraction distance. The use of tractors to extract timber also fell from use. In this system, tractors (often agricultural or industrial ones) were used to haul timber on sleds over ice roads, usually to a river-side landing for floating. Again, this method was really only suitable for fairly short distances and so increases in extraction distance (along with the increases in the power of the available logging lorries and the demand to de-seasonalise the logging process) led to the method's decline.

The two dominant methods are, and have been, narrow-gauge railways and lorries. Some ordinary-gauge railways were built, as has been mentioned, but these were only used for very large-scale extraction to the very largest of lower landings - for example, where at least half a million cubic metres were processed annually (Koroleff, 1959). However, the importance of railways began to decline relatively in 1960 and by 1972 it was reported that no more railways were being built for the purposes of timber extraction, although those already in existence were still being used (Krainov *et al.*, 1972). The reason for this shift was probably the large capital costs of railways compared to roads. In 1959, when the relative decline of railways began, it took between 1,000 and 1,400 man-days to build one kilometre of railway, whereas it needed only 800 to 1,000 man-days to build the same distance of gravelled round-the-year road. Also, the development of logging lorries and roading technology saw a decline in the specific costs of this sort of extraction. In the 1950s the average load per trip for a logging truck was about 5 to 10 tonnes - more if the lorry was hauling sleds on an ice road; now, typically, 15 to 23 tonne loads are the norm.

The power source used in timber extracting lorries had a similar history to that exhibited by logging tractors. To make forestry self-sufficient, lorries powered by wood-burning gas producers and by steam were developed. For some time it seemed that these types of lorry were to find widespread use (Korotonoshka and Shebalin, 1950). In the late 1950s about 35 per cent of logging trucks were powered by wood-burning gas producers and although steam power accounted for only a very small proportion of all lorries, some authors were enthusiastic about them.

However, as the supply of petrol and diesel fuel improved these more exotically powered machines began to be used mainly in the more remote logging areas where liquid fuel was difficult to obtain. Eventually, as the development of diesel powered engines improved over the other types of power sources, the same thing happened to logging lorries that happened to skidders; now all logging lorries are diesel powered.

Lower Landing Work

As the mechanisation of logging operations has progressed one effect has been to relocate more of the work from the forest floor to the lower landing. For example, cross-cutting and delimbing operations have been gradually moved to the lower landing from the forest floor or upper landing. However, in Table 4.2, of the total number of man-days required to produce 1,000 cubic metres, the overall percentage needed for work at the lower landing has remained roughly constant, showing only a slight increase from 19.4 per cent in 1963 to 21.4 in 1970. At the same time there has been an increase in the use of the full tree and tree length systems of extraction. Therefore, the investment in equipment at lower landings must have been sufficient in both quantity and quality to increase the productivity of labour at a rate high enough to overcome this increase in the amount of work to be undertaken at the lower landing.

However, unlike logging work, it is impossible to generalise about the actual amount of work carried out at the different types of lower landing. The requirements for manpower can vary greatly depending on the type of landing being considered. For example, a lower landing situated on a railway producing a large variety of products such as pit props, pallets, etc., as well as a range of cross-cut log lengths in an area where a full tree or whole tree extraction system is used would require equipment for loading, unloading, moving, sorting, debarking, delimbing, sawing, splitting and assembling. The proportion of the total amount of work required to produce 1,000 cubic metres of timber needed in operations at the lower landing might be as high as 75 per cent. But the figure would be much lower at a small landing on a river-side where the timber is floated loose.

The gradual mechanisation of work at the lower landing is far less interesting a story than that of the mechanisation of logging. The conditions encountered at lower landings are far less restrictive than those found on the forest floor. Therefore the mechanisation of the various operations followed a far easier course. The fact that lower landings are at least of a semi-permanent nature meant that efficient electrical generating equipment could be installed, or if the lower landing was large enough and situated near to other populated areas the State power system could supply reliable electricity. So none of the problems that were encountered in the use of electrically powered logging equipment were common at lower landings. Electric cranes, saws, green chains, conveyors etc. appeared and stayed. The only competition from diesel power comes from diesel-engined loaders and transporters which are mobile and can be used anywhere in the lower landing.

After the war much of the on- and off-loading of timber was carried out by hand and so one of the first steps was to mechanise this labour-intensive operation. Large cranes were introduced and the production of cranes for use at lower landings more than doubled in the first few years after the war (Popov, 1948). It was hoped that this would very quickly double the productivity of labour from its low post-war level of about 5 cubic metres per man-day (Voronitsin, 1957). However, the figures in Table 4.2 show that in 1963 it still required 127.7 man-days to produce 1,000 cubic metres - a productivity of labour of about 8 cubic metres. Thus the investment did not live up to expectations; part of the reason for this was that the new cranes still needed a fairly large labour input. For example, the cranes in common use now need up to 5 slingers when loading timber.

Electric conveyors and green chains were introduced tc minimise the amount of handling, although like the cranes used, large labour inputs are still required to go with the new equipment. The electric saws used at the smaller landings for cross-cutting and delimbing are developments of the types of saws used on the forest floor in the past. Their use at lower landings does not present the same problem of moving heavy cables around which was a major disadvantage of their use for felling. Also such saws are able to perform at landings without the need for increased personnel to operate special generating equipment.

Moves to continue the mechanisation of work at lower landings were held up to a certain extent by economic considerations. In 1966 *Giprolestrans* produced figures which showed that the most economic technology for a lower landing with an annual throughput of 100 thousand cubic metres or less made use of electrically powered hand-held saws and other similar low levels of mechanisation. Still, the average size of lower landings is such that, economically, there is no great incentive to increase the level of mechanisation. However, partly as a result of the concentration of production that is acting to increase the average landing size and partly because of the labour problems being faced by the forest industries, plans for the total mechanisation of work at the lower landing are in being. These will be discussed in the next chapter.

PREPARATORY AND AUXILIARY OPERATIONS

The figures in Table 4.1 show that the proportion of man-days in preparatory and auxiliary operations relative to basic operations has increased since 1950. At the same time, the productivity of labour seems also to have increased, only 219 man-days being needed in such operations to produce 1,000 cubic metres in 1975 compared to 540 in 1950. However, the data in Tables 4.1 and 4.2 show that most of this reduction occurred before 1963 and that the rate of productivity improvement has probably shown a similar trend to that shown by basic operations - a rapid improvement in the late 1940s and early 1950s followed by a less impressive performance in the 1960s and 1970s.

Preparatory and auxiliary operations (preparing the felling area, maintenance and repair work, roading, etc.,) clearly benefited from the general improvement in equipment experienced during that time. For example, preparing the felling area by felling dangerous trees and preparing loading areas and roads is now easier with the introduction of chain saws and crawler bulldozers with suitable attachments. In this area one would expect that productivity would show a similar trend to that of basic operations. The introduction of electric saws and wood powered tractors and electric winches improved productivity but a large surge in productivity occurred when the less reliable form of technology was replaced by more conventional and reliable types of equipment.

Of course, the use of more and more machinery has necessitated an increase in personnel for maintenance and repair; this tended to offset the increase in productivity that the new equipment brought. To begin with this effect was compensated by a reduction in the time needed to care for and feed the horses used in skidding and extraction. Also the notorious problems of obtaining spare parts for any machinery in the Soviet Union has meant that, with the introduction of more capital, each *lespromkhoz* has felt the need to have its own workshop and skilled workers to produce many of the spare parts for the equipment. This problem is reducing the impact of many of the new machines now being introduced and this will be discussed in the next chapter.

ORGANIZATIONAL CHANGES

Varaksin and Stupnev (1974) are some of the few Soviet experts to analyse in any depth the pattern of labour productivity and levels of capital in logging through time. Apart from the development and introduction of new types of equipment, they note that changes in labour organisation have been made possible by the new machinery and that these have also affected levels of labour productivity. In fact, they attribute the rapid growth of productivity shown in the later 1950s almost entirely to 'the massive movement towards more advanced forms of organisation of production ... and a better use of the equipment already in existence in logging concerns'. The period they are referring to roughly corresponds to the period 1950 to 1963 mentioned in the introduction which saw an increase in labour productivity of a little more than 8 per cent per annum.

There were two main labour force re-organisations in the 1950s and 1960s. First, fairly large groupings of workers were formed - up to twenty-four workers per group, supposedly well integrated and headed by a foreman. Each member had a specialisation - felling, tractor driving, etc., - and work was organised so as to prevent hold-ups. For example, fellers worked one shift 'ahead' of the delimbers who were three or four shifts ahead of the tractor drivers and chokermen. Others worked at preparing the area for the main group and helping in loading operations. The actual details, of course, varied with the local conditions but the system, when it was introduced in about 1953, increased output by between 10 and 20 per cent according to some reports.

However, by the late 1950s this system was falling from favour. With such large groups maintaining high output levels from every member proved to be impossible. It was possible to 'hide' in the size of the group. Also, with each member of the group having his own specialisation, there was no flexibility for absences or illness. As a result of this disenchantment the small complex brigade was introduced - *malaya kompleksnaya brigada*. This is still the main form of labour organisation and it typically consists of small groups of only three to eight workers - more if the tree length system is used, less if the full tree system is in use. Each member has to have more than one area of expertise and must be prepared to change his job to suit the day-to-day needs of the brigade and logging concern as a whole. This increases the flexibility of the system and the smaller size of the unit has made the individual's contribution to output more apparent and has meant that there is less slack in the grouping.

Other changes were also occurring which Varaksin and Stupnev claim helped to increase labour productivity. A more rational approach to logging saw the increase in the use of stock-piles between the various operations in the logging process which helped to reduce the inter-dependence of the workers doing different tasks and so helped to minimise wasted time.

In fact in the post-war period under discussion, the whole approach to logging has been to tackle organisational problems in a more rational manner. A new subject, work study (*nauchnaya organizatsia truda*) was introduced and departments in the forestry institutions were formed to teach and research the subject. This subject and these departments now help to advise on the best way to use the new equipment coming into use.

CONCLUDING REMARKS

The history of the development of logging technology is rather important for the future of the industry. In the introduction to this chapter it was shown that the growth in labour productivity was best seen as consisting of two stages, a period of rapid growth until the early 1960s followed by a period of slower growth. When this was related to the input of capital, measured in the form of horse-power per worker, it was seen that a lagged relationship existed, with increases in labour productivity occurring up to four years after the introduction of new equipment.

This is a view with which most Soviet experts would agree. Varaksin and Stupnev (1974) claim that only a small increase in the productivity of labour resulted directly from the initial introduction of new equipment such as the KT-12, electric chain saws and the beginning of the mechanisation of loading and unloading. They state:

> 'The quantitative change (in the amount of capital equipment) ... did not lead to any notable growth in the productivity of labour ... but rather prepared the ground for considerable qualitative changes in logging production.'

They then go on to make the statement quoted earlier, that a large part of the subsequent growth in productivity was attributable to organisational changes in the logging and conversion process.

The implications of this position are clear. According to Varaksin and Stupnev, any new equipment does not of itself greatly increase labour productivity, but rather the learning effects and the re-organisation that the new technology permits (which follow the introduction of the machines after a period of time) are what really affects output. Thus, the new equipment now being introduced will begin to have a significant effect on output only if modern methods of using the new technology are evolved. Therefore, the areas where the potentialities of new equipment become evident are not in the specifications and performance of the machines themselves (although that is important) but in the various ways in which they may be used. Such an interpretation is of vital importance at the present time as new feller-bunchers, feller-skidders, grapple skidders, etc., are coming into use. It would suggest that the present sluggish rate of growth of labour productivity would not suddenly increase with the increased use of the new machines but only after a period of gaining experience in their use and after organisational changes had been made.

Thus the disappointing growth in productivity from 500 to 530 cubic metres per annum between 1970 and 1975 (see Fig. 4.1) was to have been expected despite the beginnings of the introduction of the new equipment. Maybe by the late 1970s or early 1980s, as new methods evolved, the productivity of labour would take off again and show a similar rate of growth to the 8.1 per cent that it showed between 1950 and 1963.

However, two points throw doubt on this view and suggest that it is optimistic. First, organisational changes, especially where they concern the labour force, tend to produce short-term gains in productivity. The changes from *ad hoc* organisations of workers to large groupings of twenty to twenty-four workers and then to small, complex brigades led to immediate increases in productivity in the groups studied. But no work that the author is aware of shows the long-term effects of such re-organisations. It is possible that, after the 'novelty' effect of a re-organisation has worn off the initial increase in productivity is lost and the rate of productivity growth falls back to its former low levels. It is known, for instance, that the output of workers whose work is being shown an interest in, is greater whilst that interest is exhibited than when the management withdraws its attention. If this is true for the Soviet experience then one of the reasons for the disenchantment with the larger groupings of workers that was beginning to become apparent in the late 1950s may have been that productivity was not growing at the rates promised by the early experiments. Now a similar disenchantment is being expressed with the small brigades and larger groups are again being advocated on the basis of fairly short-term trials. The reason given for the change is that larger brigades will be able to exploit the potentialities of the new systems of machines. If, however, this short-term effect does characterise the path of productivity changes through time in response to organisational changes, new enlarged brigades will not have the desired long-term effect of allowing a continued increase in the rate of growth of labour productivity.

Secondly, the lagged relationship between changes in productivity and changes in capital intensity may be better explained by examining the nature of the equipment introduced into logging. The massive post-war investment in logging machinery started off in what turned out to be the wrong direction. Trying to make the forest 'log itself' by a self-sufficiency in its energy needs led to the introduction of machines which are no longer in use, and neither are derivatives of them - gas powered tractors, electric felling saws, steam lorries, portable wood-burning generators, electric skidders, etc. Thus the line in Fig. 4.1, expressing the number of horse-power per worker, conceals the nature of the path taken by investment in logging equipment.

In the period to the mid-1950s the KT-12 was replaced by the TDT-40; although the power of the two tractors was similar, and so would show no large change in the number of horse-power per worker, there would have been a marked effect on productivity. The diesel-engined tractor was 30 to 50 per cent more productive and would have consumed fewer man-hours in fuelling up. As well as that, the maintenance of the more conventional TDT-40 required less time so it would spend more time on the job.

The move away from electric power to petrol and diesel made the various elements of the logging process more independent. The reliability of the whole logging operation was increased with the removal of the critical generator but this was not associated with an increase in the number of horse-power per worker.

Finally, the labour laws in the Soviet Union will mean that any redundancies usually have to be absorbed by natural wastage. While this is not a slow process in Soviet forestry, it would mean that new technology would only 'show up' in the reported productivity figures for the industry as a whole after a suitable time period.

The argument above, in the absence of good data, is naturally speculative. However, it does seem that productivity increases in the past can be better explained by the nature of the capital equipment introduced rather than by a 'learning effect' producing a lagged response in organisational factors - the reverse of the main thrust of the argument presented by Varaksin and Stupnev (1974) and other Soviet experts. Thus, the implication is that the performance of new machines in increasing labour productivity should be fairly immediate - unless good technical reasons can be found otherwise. Whilst it is no doubt true that the potentialities of a new system of machinery will not be entirely apparent in the first few years of its use, the major part of the benefits that new systems bring should be registered fairly quickly. In the light of this conclusion, we can now direct our attention to the present and offer some tentative predictions concerning future increases in labour productivity that the planned developments in Soviet logging technology will bring. This will be discussed in the next chapter and, when combined with the analysis of the supply of labour to forestry presented in Chapter 6, will provide the basis for forecasts of timber output from the Soviet Union's forests.

5
The Possibility of Future Productivity Increases

INTRODUCTION

In the previous chapter we examined the evolution of the logging system that was in most common use in the Soviet Union in the 1970s. That was the combination of a petrol-engined chain saw, winch and choker equipped crawler tractor skidding the felled trees over a relatively short distance, axes to delimb the trunks and the extraction of the logs by lorry to a lower landing where varying degrees of mechanisation were employed in log breakdown. However, it was seen that this system and its development had not achieved the success hoped for in raising productivity; in fact the progress in mechanising the logging industry has made some people's work more and not less arduous. For example, as the power of the tractors employed has increased, from the 42 horse-power TDT-40 to the 110 horse-power TT-4, so the daily work norm has increased. In the late 1940s the daily work norm was about 40 cubic metres per tractor per shift; in 1970 it was 70. This meant that the amount of wood that the chokerman had to gather to the cable system of the skidder had nearly doubled. This point will be discussed again in Chapter 8 as an example of the failure of Soviet investment appraisal techniques.

The disappointing performance that the recent mechanisation of Soviet logging has shown in raising the productivity of labour has, of course, been noted by Soviet authors; one of the reasons that they have put forward for it is illustrated by a distinction which they draw between the mechanisation of labour and the mechanisation of the process. This distinction is illustrated in Table 5.1. The mechanisation of the process (*mekhanizatsia rabot*) describes the volume of the various operations that make up the process that are carried out with the aid of some type of machine or mechanism. As can be seen in 1971 felling, skidding, loading and unloading at the upper and lower landings, extraction and cross-cutting into assortments at the lower landing was carried out totally by some means of mechanisation and almost 30 per cent of all sorting and stacking of timber at the lower landings was carried out entirely by hand, whilst only 9.6 per cent of all delimbing was performed with the aid of some sort of mechanical equipment.

However, the mechanisation of labour (*mekhanizatsia truda*) is less than the figure for the mechanisation of the process. The mechanisation of labour is an index in common use in the Soviet Union.

TABLE 5.1 Mechanisation Levels in Logging Operations in 1971

Operation	Mechanisation Level of the process %	Mechanisation Level of labour %
Felling	100	67.0
Delimbing	9.6	3.2
Skidding	100	54.1
Loading at upper landing	100	37.7
Extraction	100	90.0
Unloading at lower landing	100	40.0
Cross-cutting	100	75.0
Sorting	70.9	8.0
Stacking	70.7	35.5
Loading assortments at lower landings	100	39.5

Source: Varaksin and Stupnev, 1974

It expresses the ratio of the number of man-hours in a process that are mechanised to the total number of man-hours in that process. This is shown in equation 1. Here ML is the index of mechanisation of labour, m_1 is the number of man-hours of the process carried out with the aid of some sort of equipment or mechanism, and m_2 is the number of man-hours in the process that require manual labour.

$$ML = \frac{m_1}{m_1 + m_2} \cdot 100\% \qquad (1)$$

Thus, to take skidding as an example, although the operation is totally mechanised, that is, there are no skidding operations that rely totally on muscle power, only 54.1 per cent of all the man-hours required to skid the timber felled are carried out with the aid of mechanisation. Therefore 45.9 per cent of the total number of man-hours are employed at manual labour. This shows the heavy requirement in that process for manual labour in connecting the logs to the choker system, nearly one-half of the number of man-hours used in skidding being used here.

The rest of the figures show a similar pattern. All of the operations, despite being mechanised to a certain extent, still require considerable inputs of manual labour. Felling still requires a lot of manual labour although the actual act of severing tree from stump is mechanised. Much of the loading and unloading at the upper and lower landings, sorting and stacking require heavy physical exertion as part of the process.

Thus, much of the input of capital into the logging process has been of the form that still needs the combination with it of a large amount of manual labour. Man-handling of timber is still an important part of many of the operations needed to get the timber from the stump to the stock-piles of the lower landing, especially in the areas of loading, sorting and stacking. Timber has to be handled on and off the equipment. This means that investment in more powerful models of the equipment in use leads to the same position as has been the case with skidders. As long as capital equipment is combined with substantial amounts of hard, physical labour, any increase in power or the rate of production of the piece of equipment in question will necessarily result in an increase in the input of manual labour that the act of capital investment was, ostensibly, attempting to reduce. Therefore the very direction of investment in the logging industry is now being questioned by some Soviet experts. What is needed, they argue, is the full and complete mechanisation of the processes that constitute the whole logging and preliminary log breakdown cycle.

THE SYSTEMS PLANNED

In the absence of freely moving prices and a market system to decide on the 'optimum' degree and type of mechanisation in logging - a problem examined in Chapter 8 - it is perhaps not surprising that this subject is treated in a much more 'technical' way than it would be in the West. Before looking at the arguments put forward by the Soviet experts, it might help to begin with a few generalisations.

Logging systems can consist, reasonably, of up to nine different operations. These are: felling, delimbing, cross-cutting, chipping, bunching of stems or logs, skidding, forwarding (the skidding of cross-cut lengths), loading at the upper landing and extracting to the lower landing. The permutations that can be worked into the mechanisation of these nine basic operations are very large. At one extreme one machine could be used to fell the tree and then carry it to the lower landing, or, alternatively, separate machines could be used for each of the operations. Also, there is a variety of possible locations for many of the constituent operations of logging. For example, delimbing of the felled tree could occur immediately after felling on the forest floor, or after skidding at the upper landing, or, if the whole tree extraction system was to be used, it could be deferred until the whole tree arrived at the lower landing. Various factors would affect this choice; if there was a high demand for chips or low quality wood at the lower landing, delimbing there would be best. If there was no demand whatsoever for low quality wood, delimbing on the forest floor might be best as this would avoid unnecessary skidding of unwanted branches and would prevent the accumulation of stock-piles of brash at the upper or lower landings. However, as delimbing at the upper landing would probably be easier, technologically or mechanically to organise, a good compromise in this situation might be delimbing at the upper landing - only transporting unwanted branches and tops a few hundred metres and accumulating only small, easily disposable, piles of brash at the upper landings.

Similarly, the cross-cutting of logs into the required bolt lengths can be carried out at different places. Cross-cutting at the lower landing would probably be more accurate and less wasteful in that better cutting patterns could be adopted than would be available to someone working in the conditions of the forest floor. Also, conditions for the workers themselves would be better at the lower landing rather than on the forest floor or upper landing. However, tree tops, etc., at the lower landing would represent wasted energy in the transport of unwanted material to the lower landing. The location of cross-cutting and delimbing is the basis for one of the common distinctions between various types of harvesting systems. In the short wood system cross-cutting and delimbing are carried out on the forest floor; in the tree length and whole tree systems cross-cutting is performed at the lower landing with delimbing being performed at the upper and lower landings respectively.

This is one area where a choice has been made which, at least in part, improves working conditions. There is a greater use of the whole tree and tree length logging systems in the Soviet Union than in the West (see Table 5.2) and it is intended to increase their relative use. In 1990 it is hoped that over one-half of all timber logged will be extracted as whole trees. This tendency to move operations from the forest floor and upper landing to the lower landing is coming about partly as a response to the increase in demand for low quality wood, branches and tops for particle board and paper manufacture. This demand is the result of the problem of increasing logging output. If more of the tree is converted into some sort of final product the productivity of labour, expressed in terms of roubles of final product rather than cubic metres of timber felled, will be that much greater.

TABLE 5.2 Uses of the Whole Tree, Tree Length and Short Wood Logging Systems

	% By Volume Logged by the Main Ministry			Use in Europe*
	1972	1980	1990	1974
Whole tree system	5 - 7	20 - 25	50 - 55	1
Tree length system	68 - 75	60 - 70	30 - 40	47
Short wood system	20 - 25	10 - 15	10 - 15	52

* European figures are defined slightly differently to Soviet data. They understate the prevalence of what is defined here as the short wood system.

Source: *Minlesprom*, 1972 and Loffler, 1978

If the whole of the logging system is to be mechanised then one or a whole system of machines will be needed to fulfil the various constituent operations. Varaksin and Stupnev (1974) note three major features that any machine should possess. First there should be 'homogeneity' of the object being worked with. That is, a machine should operate entirely with individual trees, or bunches of trees, etc., and not sometimes with the one and other times with the other. Secondly, there should be independence of operation; that is, a machine in a system should not be dependent on the fulfilment of operations of another machine in the system. This requirement would reduce idle machine time. Otherwise large stock-piles between machines would have to be created and this would increase the amount of handling required of the various machines in a system and so reduces overall productivity, or else would necessitate duplication of machines in the system (one feller-buncher and two delimbers, for example) and thereby increase capital investment disproportionately. Thirdly, there should be as far as possible coincidence of operation. That is, any machine fulfilling more than one operation should carry them out coincidentally and not consecutively. There should, ideally, be no periodic idle time for any of the devices on a multi-operation machine. As a further point, systems of machines should have the same base unit, for example they should be the derivatives of the same tractor unit; this would ensure that spare parts and repair problems are minimised.

In the most common logging system now in use, loading has been separated from the skidding and the extraction processes with the introduction of loaders. However, the skidder is usually equipped with a choker system of log gathering, and so is open to the criticism that it gathers individual trees and skids bunches of trees. Also, these two operations are, almost of necessity, carried out consecutively and not coincidentally. This is an interesting point for there are now being introduced many grapple skidders equipped with hydraulically powered grapples to gather trees together on the bunk of the machine and skid the bunches so formed to the upper landing. The grab is controlled by the tractor driver from inside the cab and, in doing away with the chokerman, at once increases the productivity of labour and totally mechanises the skidding process. However, because of the objections raised to this system that it gathers individual trees but skids bunches, the use of such grapple skidders is 'only an intervening stage (in technical development), a temporary, technical expedient' (Varaksin and Stupnev, 1974).

Clearly felling must be carried out with individual trees in whatever spatial arrangement that occurs, but a feller-buncher can fulfil the third requirement put forward by Varaksin and Stupnev by forming a bunch of stems as part of the felling process. Bunches of trees are left on the forest floor after they have reached a certain size and are collected by a skidder; usually this will be a grapple skidder with just one large grapple to collect the bunch of stems in one go and return it to the upper landing thereby not falling foul of the argument above. There timber will be loaded and possibly delimbed by separate machines. Here, if the technical characteristics, especially output of the various elements of the system, are compatible there is little wasted machine time and the machines spend the maximum time doing what they were designed for.

Using a feller-skidder rather than a feller-buncher and separate skidder has the economic advantage that one fewer machine is needed (along with its operator) but means that, for a similar felling capability, overall productivity is reduced as the feller-skidder will spend much of its time skidding timber when, if there were a separate skidder it could be felling more timber. The feller-skidder does not pass the third of the conditions laid out above - there are consistently periods of time when it is skidding when its felling attachments will not be in use. However, in areas with low skidding distances it might be a better machine to use. Much research has gone into attempts to discover which of these systems is superior.

In the West, a delimbing attachment is often included in the felling apparatus of a feller-buncher or feller-skidder and the tree is severed at the base, delimbed and placed in the bunk of the machine for bunching or skidding. This system might be objected to on the grounds of the third criterion put forward earlier - felling, delimbing and bunching or skidding are carried out consecutively. In several Western machines the delimbing of the felled tree is carried out as the manipulator turns the tree towards the bunk and prepares to fell the next tree. But in any case, for one reason or another, such feller-delimber-bunchers/skidders are only now beginning to appear in experimental form in the Soviet Union.

Also tree harvesters are popular in the West, especially in Scandinavia where the short wood system is used. This type of machine, which fells, delimbs, cross-cuts and forwards the logs to the upper landing, would seem, except in special circumstances to go against the overall philosophy of logging operations in the Soviet Union as it maximises the number of operations performed on the forest floor and does not allow the use of branches and low quality wood for conversion into chips at the upper or lower landing. It might be a good system to employ, however, in forests near pulp and paper mills where bolt lengths can be produced and transferred easily to the mill.

The Projected Logging Systems

TsNIIME, its affiliated institutes in the Caucasus and Irkutsk, *SibNIILP*, *CNIILP*, *SevNIILP*, *KarNIILP*, the Moscow and Ural Forest Technical Institutes, the Kirov Forestry Academy and the Siberian Technical Institute worked out a plan for the mechanisation of logging in the forest industries to run until 1990. This was scrutinised by the members of the Directorate of *TsNIIME* and by members of the Ministry and was agreed by the ministry concerned with the construction of tractors and other machines. It was published on 14 November 1972 as *Prikaz* No.337 entitled 'On the Creation and Introduction of System of Machines and Equipment for the Complete Mechanisation of Production in the Forest Industry for the Period Until 1990' (*Minlesprom*, 1972).

The aim of the plan was to substitute high productivity machines for manual labour whilst keeping capital costs to a minimum. The plan divided the 'future' into two periods; in the first, until 1980, modified machines based on already existing equipment were to be introduced. This, it was hoped, would replace quickly all manual labour in the various processes as cheaply as possible. The second period, from 1980 to 1990, was to be characterised by the introduction of systems of new machines, also designed to exclude manual labour from logging but representing an optimum technical and economic set of systems and equipment. The introduction of these machines, it was hoped, would bring two other results apart from the exclusion of manual labour from logging. The machines were expected to reduce the amount of preparatory work in forests before the main logging operations (felling dangerous trees, etc.,) and industrial injuries would be reduced as a result of safer working conditions.

The plan is shown in Table 5.3. The percentage usages shown in the table indicate the amount of all the volume logged by *Minlesbumprom* that the various systems were to account for. Because maximum and minimum figures are shown for these values, they do not sum to 100 per cent. In the table the systems are referred to with the same 'index' that is used in the Soviet publications. Thus system L1s is logging system number one to be used with small-sized trees, and so on. The delimbing phase, where it is explicitly mentioned, is optional in the sense that it may be excluded if the whole tree system were to be used for extraction. In this plan it was projected that the whole tree system would increase at the expense of the tree length and short wood systems. The figures projected are those that were given in Table 5.2.

Systems L1s and L1l were the systems that were most common in the recent past and it is seen that their use is to be phased out in order to reduce the manual labour involved in chokering the load for skidding. As was noted earlier, the mechanisation of the chokerman's work by the use of grapple skidders with separate hydraulic arms that picked up individual stems that had been felled by chain saw, systems L2s and L2l, suffered from the theoretical deficiency that the skidder did not operate with a 'homogeneous' object and was thought to represent only a temporary expedient. This is shown by the projected usage of these two systems. Whilst they are ideal for the first part of the plan's period, until 1980, as all the work in the logging process would have been mechanised, they do not represent optimum systems and so their use, after rising to between 33 and 37 per cent in 1980, was to decline into the 1980s.

The most popular system in this plan is L3, and L3s and L3l were to account for between 29 and 34 per cent of all logging operations in 1990. The systems involving the use of a feller-skidder, criticised for their partial use of the felling attachment on the feller-skidder, are present in Table 5.3 as systems L411 and L5. Here, presumably, they were being suggested for use in situations when the skidding distance was relatively short. System L412, where a high productivity felling machine is to replace chain saws, was to be used where tree sizes and machine productivity were expected to be large enough to justify this. The incorporation of chip production into a logging system, with the chips to be produced at the upper landing or on the forest floor, is present in system L6 but will only account for between 2 and 3 per cent of output in 1990. Most chip production will be located at the lower landing. Tree harvesters, popular in the West, will account for only 5 to 10 per cent of trees felled in 1990 (system L8) and will be used in areas near large wood processing concerns.

The Present Performance of the Projected Systems

The systems presented in Table 5.3 are now beginning to come into use. If the plan had been fulfilled, productivity in felling and skidding should have increased dramatically. According to Table 4.2 in the previous chapter, 158.8 man-days were required to fell and skid 1,000 cubic metres of timber in 1970 in the *Archangelsklesprom*

TABLE 5.3 Projected Systems of Logging Until 1990

Index	Description of system	No. of machines	No. of workers	Productivity /shift over-all	Productivity /shift of labour	Area of use	Projected use % 1980	Projected use % 1990
L1s	Chain saw Crawler tractor with winch and and choker system - TDT-55 Manual delimbing	1 1 -	1 2 3	50	$8\frac{1}{2}$	North-west, north and central European parts of Russia Small average tree sizes	11-13	2-3
L1l	Chain saw Crawler tractor with winch and choker system - TDT-75 or TT-4 Manual delimbing	1 1 -	1 2 3	75	$12\frac{1}{2}$	Urals, Siberia, Far East Average and large tree sizes	15-17	2-3
L2s	Chain saw Grapple skidder - TB-1 Delimbing by special chain saw - BS-1	1 1 2	1 1 2	65	$16\frac{1}{4}$	North-west, north and central European parts of Russia Small average tree sizes.	15-17	6-8
L2l	Chain saw Grapple skidder - LP-18 Delimbing by special chain saw - BS-1	1 1 2	1 1 2	80	20	Urals, Siberia, Far East Average and large tree sizes	18-20	6-7
L3s	Feller-buncher Wheeled or crawler grapple skidder Delimbing machine at upper landing	1 1 1	1 1 1	150	50	North-west and central European part of Russia. Clear-felling and thinning in Group I forests. Small average tree sizes.	5-6	16-19

TABLE 5.3 continued

L31	Feller-buncher Wheeled or crawler grapple skidder Delimbing machine at upper landing	1 1 1	1 1 1	200	$66\frac{2}{3}$	Urals, Siberia, Far East. Clear-felling and thinning in Group I forests. Average and large tree sizes	3-4	13-15
L411	Feller-skidder Delimbing machine at upper landing	2 1	2 1	200	$66\frac{2}{3}$	Urals, Siberia, Far East. Large average tree sizes and poor soil characteristics	12-18	14-20
L412	Felling machine Grapple skidder Delimbing machine at upper landing	1 2 1	1 2 1	300	75	Urals, Siberia, Far East. Large average tree sizes and poor soil characteristics	4-6	12-16
L5	Feller-skidder Delimbing, cross-cutting at upper landing	1 1	1 1	100	50	Areas serving large wood processing concerns	up to 2	4-5
L6	Feller-buncher Grapple skidder Chipper at upper landing	2 1 1	2 1 1	150	$37\frac{1}{2}$	Areas with small average tree sizes serving paper producers.	-	2-3
L7	Chain saw Winch skidding system	1 1	1 3	60	15	Mountainous areas of Siberia, Caucasus, for thinning and clear-felling	2-4	4-6
L8	Tree harvester to fell, delimb, cross-cut and skid the timber to the upper landing	1	1	50	50	Forests serving large wood processing concerns	-	5-10

Source: *Minlesprom*, 1972

region. Allowing for the lower-than-average productivity in this area, this would suggest an overall productivity of labour in these operations of about 7 cubic metres per shift. Using the formula derived in the Appendix to this chapter, the plan presented in Table 5.3 implies a productivity of labour for felling and skidding in 1980 of 19 cubic metres per man per shift and, in 1990, of about 35. This represents an annual growth rate in productivity of 10.5 per cent from 1970 to 1980 and 6.3 per cent from 1980 to the end of the plan period in 1990.

However, the investment programme up to 1980 has not been met. From the figures that are available the following emerges. Consider Tables 5.4 and 5.5. In Table 5.4 certain operations are highlighted in the various systems described in the plan and in Table 5.3. These are felling by machine other than chain saw, skidding by some form of grapple skidder and delimbing by some form of machine other than chain saw. These are picked out because they are the most often

TABLE 5.4 Projected Mechanisation Levels

System index	Average projected use 1980	1990	Machine felling	Skidding with grapple skidder	Mechanical delimbing
L1s	12	2½			
L11	16	2½			
L2s	16	7		*	
L21	19	6½		*	
L3s	5½	17½	*	*	*
L31	3½	14	*	*	*
L411	15	17	*		*
L412	5	14	*	*	*
L5	1	4½	*		*
L6	0	2½	*	*	*
L7	3	5			
L8	0	7½	*		

Average projected use is the average of the two extreme figures for the usage of the various systems shown in Table 5.3. The totals do not therefore equal 100 per cent.

referred to in the Soviet journals. After identifying these operations it is easy to calculate their future occurrence by summing the figures for the use of each of the individual systems. The figures arrived at are shown in Table 5.5. The 1980 figures are the most important ones in judging the performance of the logging industry and they have been repeated often outside of the immediate context of the *Prikaz* No.337. In a 1976 article the volume of timber to be felled by felling machines of some description in 1980 was predicted to be between 75 and 80 million cubic metres of timber or about 35

TABLE 5.5 Projected Mechanisation Levels

	Projected percentage of all volume felled 1980	1990
Machine felling	30	77
Skidding with grapple skidder	49	61½
Mechanical delimbing by delimbing machine at the upper landing*	23	33

* Calculated on the assumption in the 1972 plan that extraction of whole trees to the lower landing would account for 22½% and 52½% of the total volume felled in 1980 and 1990 respectively.

per cent of the total to be felled by the Ministry. The same article suggested that the proportion of timber skidded by grapple skidder would be about 30 per cent and the volume delimbed by machine about 27 per cent (Bagayev, 1976). Alternatively, the figures given in another article are 35, 45 and 27 per cent respectively for machine felling, grapple skidding and mechanical delimbing (Medvedev, 1976). In a 1977 article (Baranovski, 1977) the percentages are 36.6, 47 and 31 and as late as January 1978 (Timofeyev, 1978) 35, 'more than 50' and 26 per cent. Whilst these figures show some variation, they are not totally dissimilar and are close to the percentages in Table 5.5.

However, in 1979, there was a dramatic shortfall compared to the figures for 1980 given above - far greater than can be explained by the fact that the year was 1979 and not 1980. For mechanised felling the figure in the Ministry was 26 million cubic metres or about 12 per cent of the total, for skidding with grapple skidder the recorded amount was 48 million cubic metres or about 21 per cent and for mechanised delimbing 35 million cubic metres or about 15 per cent (Anon, 1979). In 1980 the figures were no better. They were: 29 million felled by felling machine of some kind or other, the volume of timber skidded by grapple skidder was actually down to 41 million as was the amount of mechanised delimbing, down to 31 million and in 1981 the figures remained unchanged. As a result of the shortfall in the implementation of the plan, the suggested mechanisation levels for 1985 seem to have been adjusted downwards. In 1980, the suggested figures were 60 million cubic metres of mechanised felling, 85 million cubic metres skidded by grapple skidder and 60 million cubic metres of mechanised delimbing (26, 37 and 22 per cent of the total) and in 1981 the figures had been slightly reduced to 53, 80 and 51 millions respectively (Stupnev, 1981 and Prodaivoda, 1981). So looking at Table 5.5 puts the 1972 plan for capital investment more than five years behind schedule in terms of the volume of timber being processed by these new advanced machines.

The reasons for this poor performance may be that the number of machines supplied to the logging concerns is too few, or that the machines are not as productive as was planned, or, of course, some combination of these two. To answer these questions it is useful to look at individual operations in a way similar to the previous chapter.

THE MACHINES TO BE INTRODUCED

Felling Machines, Feller-bunchers and Feller-skidders

It is worth noting at the outset that there has been a considerable time lag between the development of machines for felling and other operations and their actual introduction into the logging industry in any appreciable quantity. This is a feature common to many branches of the Soviet economy and can be partially explained by the nature of the incentive system built into the planning mechanism in the Soviet Union. Basically, there is an incentive to produce old types of products rather than new ones, the production of which might lead to the under-fulfilment of the plan. This problem will be discussed in Chapters 7 and 8.

For example, Stupnev (1979) reported that in early 1979 'more than 500 felling machines, feller-bunchers and feller-skidders were in use in the Ministry's forests'. In 1980 the figure had risen to 1,340 (*Lesnaya Promishlennost*, 5.8.80), However, the machines that form the basis for those being introduced, and in some cases the actual machines themselves, were prototypes as early as 1965 and were supposed to go into full production in 1968. In the late 1960s two basic machines were still under test; these were the LP-2 and VTM-4, designed as feller-bunchers and feller-skidders respectively.

The basic characteristics of these machines and others to be discussed in this section are given in Table 5.6.

The LP-2 feller-buncher. The LP-2 was designed by *TsNIIME* and the Latvian Scientific Research Institute for Forestry as a feller-buncher based on the TDT-55 tractor. The winch and logging plate of the TDT-55 were replaced by a separate cabin, hydraulically controlled arm with felling head using a chain saw and a bunk in which the hydraulic arm placed the felled trees for bunching and in which they were held. At one position the operator could fell between 8 and 10 trees and would then move 5 or 10 metres to another position to fell a new group of trees. After felling several trees to form a full load in the bunk, the bunch of trees so formed was deposited on the forest floor for later skidding (Bozhak, 1968 and Kushlyaev, 1969).

The planned productivity of this machine was about 60 cubic metres per machine shift; however, the machine did not really live up to expectations. To be economically effective in association with the TB-1 grapple skidder (see a later section in this chapter) between 60 and 90 cubic metres per shift had to be felled. But tests showed that this figure was very difficult to achieve. Early tests showed that the productivity of an operator using the LP-2 remained at about 30 cubic metres per shift for up to two years, rising only to the planned levels in favourable circumstances after that period (Burdin, 1973). This had the bad effect of increasing costs per cubic metre at the upper landing by up to 87 per cent in one *lespromkhoz*. Even after seven years' use in one area the LP-2

TABLE 5.6 Characteristics of Fellers, Feller-bunchers and Feller-skidders

		VTM-4	LP-2	LP-19	LP-50	VM-4	LP-17	LP-49
Designation		feller-skidder	feller-buncher	feller-buncher	feller-skidder	feller	feller-skidder	feller-skidder
Maximum diameter of tree	cm	120	40	90	65	120	65	80
Maximum extension of felling apparatus,	metres	0.85	7.5	8.0	6.0	1.12	6.0	6.5
Weight of machine	kg	16000	12000	22000	-	-	11200	16000
Tractor base		TT-4	TDT-55	special	TDT-55A	TT-4	TDT-55	TT-4
Volume of bunch formed	cubic metres	11	5 - 8	-	-	-	-	-
Price,	roubles	12000	15000	-	-	11000	12970	-

Compiled from various sources

was only realising an output of 43 cubic metres per shift - and this as late as 1975 (Chesnokova, 1977).

The reasons for this lay not in the overall design philosophy of the LP-2 but rather in one or two minor design faults. First, the operating cabin was separate from the driving cabin, thus, in order to operate the machine throughout its cycle, the driver/operator had to move from one cabin to another - not an easy or pleasant task in the depths of a Siberian winter. Also the chassis of the TDT-55 proved to be too narrow and short for the demands placed on it by the role of feller-buncher. Outriggers had to be fitted to stabilise it and these had to be lowered to allow the hydraulic arm on the machine to reach out to its full length to fell the timber and place it in the bunk safely, and then they had to be lifted again to allow the machine to move to another felling position. These defects added to the time needed to complete a work cycle, and have been removed on later machines of a similar type of design, as we shall see later.

The VTM-4 feller-skidder. In the late 1960s and early 1970s the rival to the LP-2 was the VTM-4 feller-skidder which had been built as a prototype as early as 1965 by the Altaiski Tractor Factory in collaboration with *TsNIIME*. Based on the TT-4, which was produced at the same factory, it had a different felling attachment to the LP-2. A chain saw felling attachment was mounted on a small arm low down on the side of the tractor, but this had limited movement; it could be lifted a maximum of 120 millimetres from the ground and had a horizontal movement of 850 millimetres (Perfilov and Stupnev, 1968a). Trees would be directed to fall towards the rear of the tractor by pressure from an arm on the top of the cab. After felling the tree would be made to fall onto a large arm lying almost on the ground which lifted to transfer the stem to the bunk of the machine where it was held. When the bunk was full the bunch of trees would be skidded to the upper landing. Of course, if desired, the bunch could be left on the forest floor, using the machine as a feller-buncher.

Used as a feller-skidder, and with an average skidding distance of 120 to 200 metres, productivity per shift was expected to be between 75 and 100 cubic metres (Kanevski and Voronitsin, 1974). However, whilst the absolute productivity of this machine tended to be higher than the LP-2 (from about 40 to 80 cubic metres per shift in the introductory and immediate post-introduction phases) in general output was lucky to reach planned levels (Perfilov, 1974a and 1974b and Perfilov and Stupnev, 1968b).

A reason for this can be seen in the nature of the operation of the machine. Because of the restricted motion of the felling apparatus, the machine had to move from tree to tree, felling only one tree at each position, an operation which took up to 17 per cent of the actual working cycle of the machine. Also, in its use as a feller-skidder, about one-third of its time was employed carrying the load to the upper landing and returning empty; on top of this, an additional 5 per cent of the work cycle was spent unloading and manoeuvering at the upper landing. This, of course, violated Stupnev and Varaksin's third criterion for a multi-operation machine. In all, the felling apparatus was bound to be idle for at least half of the work time.

A comparison of the two machines and their modifications. It is obvious that the faults of the LP-2 were more easy to overcome than those of the VTM-4. A change of chassis and a modification of the controls was all the LP-2 really required. The one advantage that the VTM-4 possessed was its capacity to fell larger trees, a maximum of 120 centimetres diameter as opposed to 40 for the LP-2. The result of this is that development is now tending to proceed along two lines.

The VTM-4, whilst still in production, is being modified into a simple felling machine without a system to gather timber or a bunk to hold the timber for skidding. It is designed for use in forests with very large stem sizes because, in order to justify the high expenditure on this machine (11,000 roubles) it must give a very high output. So far the machine, the VM-4, has not seen much use. One test gives it a rate of output of 58 cubic metres per hour of actual felling time and, with about 44 per cent of its time actually working on the forest floor, this gives about 180 cubic metres per shift (Lumanov, 1975). Even so, this is a lot less than the 300 expected in the 1972 plan in system L412 which was to use a felling machine although one report (Baranovski *et al.*, 1976) did give an output of about this figure when the VM-4 was coupled with two grapple skidders; however, this seems to be the highest figure reported. In *Krasnoyarsklesprom*, with average tree sizes of between a half and one cubic metre and level ground, output was recorded in a favourable manner as being about 200 cubic metres per shift (Shtrek, 1978).

Using the type of felling apparatus mounted on a long-reach hydraulic arm such as that found on the LP-2, new types of feller-bunchers and feller-skidders have been produced; these are the LP-17, LP-19, LP-49 and LP-50. All of these are based on chassis other than that of the TDT-55 and require no outriggers. They also have only one cabin from which the felling and driving controls can be operated, and they have larger grapples to tackle larger trees. The LP-19 is used as a feller-buncher and forms bunches on the ground with the trees gathered from one felling position. The other three machines are used as feller-bunchers or feller-skidders. Engine power has also been much increased from the 62 horse-power of the LP-2 to 130 horse-power for the LP-19 and 110 for the LP-49 (Abol *et al.*, 1980). The LP-19 is the oldest of the four machines in the mould of the LP-2; it was first produced in 1975 and by early 1978 220 machines had been produced (Baranovski *et al.*, 1976). The productivity per shift in conditions with average tree sizes of about half a cubic metre ranges from 100 to 150 cubic metres per shift. The newer LP-17, LP-49 and LP-50 have not yet had such widespread tests (Deyev, 1978; Liceman, 1978 and Shugar, 1978).

New Developments in Skidding

With the use of feller-skidders, of course, the skidding phase is incorporated in the one machine along with felling. However, as feller-bunchers appear to be more favoured than feller-skidders it would appear that the separate skidding phase, and hence the need for skidders, is to remain.

As was mentioned earlier, the ideal skidder according to Stupnev and Varaksin's criteria would operate by gathering already formed bunches of stems, not by gathering individual stems into a bunch prior to skidding. However, the chokering phase of skidding is one of the areas where the manual input into the logging process is most marked, and so, even before the introduction of feller-bunchers and feller-skidders in all areas, it was planned to introduce grapple skidders designed to gather logs felled by workers operating with chain saws, and so not lying in bunches. The systems L2s and L21 describe this idea.

The first tractor designed for this purpose was the TB-1 which was developed as early as 1963 by the Kirov Forestry Academy, Leningrad and the Onezhski Tractor Factory on the base of the TDT-55 crawler tractor. It was designed to handle small-sized timber in logging system L2s in the 1972 plan; for use in larger timber the LP-11 and LP-18 were developed. These have more powerful engines, being based on the TDT-75 and TT-4 respectively. Some technical characteristics of these machines are given in Table 5.7.

TABLE 5.7 Crawler Tractors used for Grapple Skidders

		TB-1	LP-18	LP-11
Engine power	hp	62	110	130
Weight	kg	9,900	15,300	13,000
Speed range	kph	2.5 - 10.4	2.3 - 9.8	
Average ground pressure	kg/sq cm	0.51	0.53	0.52
Maximum load per trip	cub m	8	8	8
Maximum movement of grapple,	m	5	5	5
Tractor base	TDT-55	TDT-55	TDT-75	TT-4

Compiled from various sources

However, like the feller-bunchers and feller-skidders, the introduction of grapple skidders has been slower than hoped. According to the 1972 plan as described in Table 5.5 about half of all the timber logged by *Minlesbumprom* in 1980 should have been skidded by grapple skidder, but the figure for 1980 was only a little over 41 million cubic metres (*Lesnaya Promishlennost*, 5.8.80). This would seem to be accounted for, partly, by a shortfall in the supply of these types of tractors to the *lespromkhozi*. Assuming an average productivity of about 100 cubic metres per shift for a grapple skidder, an output for *Minlesbumprom* of 235 million cubic metres and 300 machine shifts per annum with no reserve machines, the 1972 plan would necessitate between 7,500 and 8,000 tractors for use in 1980.

Even allowing for the 'back-of-the-envelope' nature of the calculations, the number of grapple skidders supplied is far below these figures. The number of TB-1 tractors in use in the years 1970 to 1975 were respectively 7, 13, 84, 264 and 559 (Baranovski *et al.*, 1976). In 1978 the number of grapple skidders of all types in use was 1,860 in 1979 about 2,000 and 3,400 in 1980 (Stupnev, 1979; Timofeyev, 1978 and *Lesnaya Promishlennost* 5.8.80). Whilst these figures show a healthy acceleration of supply, they do highlight again the length of time between the development of a machine and its large-scale introduction.

The effect of using a grapple skidder is, first, to double the productivity of labour because the number of workers required is halved. Any other changes in output depend upon the productivity of the actual machine itself. The output per machine shift of these skidders is usually quite respectable; output for the TB-1 is usually reported to be in the region of 60 cubic metres per shift (Chemyanin, 1975; Perfilov, 1974a and Popov and Mikhaelov, 1975), and the LP-18 performs well, returning outputs of between 60 and 100 cubic metres (Deyev, 1978 and Shtrek, 1978).

Two factors are worthy of note here relating to the future development of skidders. First, new types of grapple skidder are under investigation and, secondly, wheeled skidders are undergoing development.

New types of grapple skidder. The grapple skidders described in Table 5.7 all have a grapple arm which can pick up individual trees and place them on the bunk of the machine; this, as was stated, enables them to be used with chain saw felling when the felled trees do not lie in bunches. This situation, for reasons given at the beginning of this chapter, is regarded as only being a temporary state of affairs. When felling is totally mechanised, feller-bunchers (or feller-skidders) will be used and will produce bunches of stems lying on the forest floor. To gather these bunches one large grapple is all that is needed to pick them up and hold them during skidding. This is more in the style of Western grapple skidders - and in Russian these are referred to as *podborshchiki-trelovoshchiki* as opposed to *bezchokerni trelovochni traktori* such as the TB-1.

In the middle 1970s these types of tractors were often mentioned in the journals as the more advanced form of skidder, but of late, they do not figure very prominently in the technical press. It seems that the LP-18 and LP-19 and developments along these lines are now considered to be the type of machine for the future. The grapple arm and separate bunk are being retained despite the use of feller-bunchers and the creation of already formed bunches of stems on the forest floor. One of the reasons for this is probably the fact that the retention of the arm gives the operator greater flexibility in the choice of the number of bunches to gather before returning to the upper landing. For the use of grapple skidders that gather a whole bunch and hold it in the gathering apparatus to be successful it is best if a whole load can be gathered at one position and this constrains the size of bunch that can be formed by the feller-buncher and means, that often, the skidder will be skidding loads that are below the volume that the power of the machine allows.

Wheeled skidders. As was mentioned in the previous chapter, the use of crawler tractors has been almost universal in Soviet forestry. Now the position is beginning to alter. Table 5.8 shows the characteristics of the types of wheeled skidders being used experimentally. The KTTs-1 was originally developed by *TsNIIME* in 1966, but now the main types of tractor being considered for use are the T-157 and K-703. Like the newer types of grapple skidder discussed above, these two were also enthusiastically mentioned in the technical press in the mid-1970s, but now do not receive the same coverage.

TABLE 5.8 Wheeled Skidders of Soviet Origin

		T-157	K-703	KTTs-1	T-80L	T-5L
Engine power	hp	150	220	100	60	50
Ground clearance	mm	515	530	550	500	500
Weight	kg	9,900	16,500	7,700	5,550	3,600
Speed range	kph	up to 32	3.1 - 34	4 - 31	up to 35	2.4 - 42
Maximum volume per trip	cub m	9	15	8	4	3.6

Compiled from various sources.

The reasons for this enthusiasm were associated with the speed of the wheeled machines. Referring to Table 4.5 in the previous chapter the maximum speed for the crawler tractors in use is between 10 and 11 kilometres per hour; but a wheeled skidder can easily better this, even when loaded. As a result, wheeled skidders give two attractive options to the planners; either they can be used over the same skidding distance as the relatively short distances used with the present crawler tractors, in which case their greater speed would allow them to return a higher productivity, or else they could be used over longer skidding distances, when their productivity would be reduced but the requirements to build logging tracks in the forest would be diminished and so some capital expenditure saved. The first option is one where the wheeled tractors would show less of a relative advantage as their higher speeds would be less easy to take advantage of. Fig. 5.1 gives an idea of the advantage that wheeled skidders have. Line I shows the productivity per shift of the K-703 and line II the TDT-75. In the tests from which the graph was derived both tractors were equipped with a winch and choker log gathering system. As can be seen, the K-703 was more productive even over short skidding distances. Even over a distance of 1 kilometre the K-703 was as productive as the TDT-75 over a common skidding distance of 250 metres. Also, its productivity did not fall off as rapidly with distance as did that of the crawler tractor. At a skidding distance of 6 kilometres it was

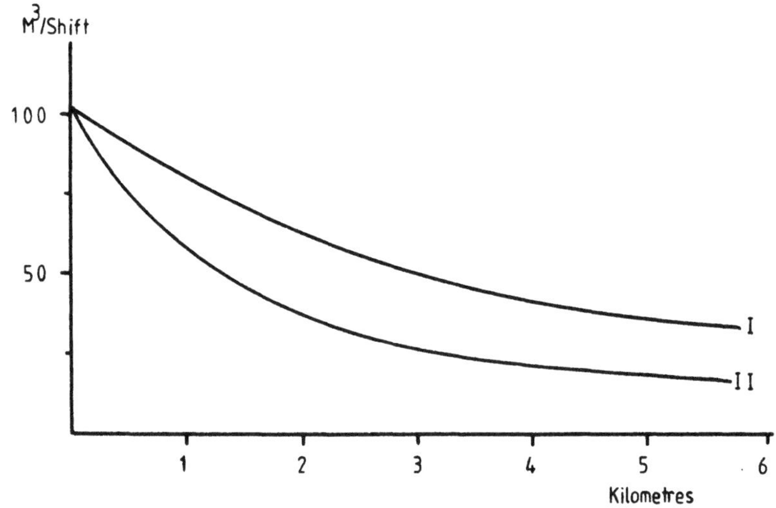

FIG. 5.1 Productivities of Wheeled and Crawler Skidders

only half as productive as the TDT-75 at 250 metres.

It is here that the possibility of reducing the roading density is apparent. As was described in Chapter 3, logging in the Soviet Union has much road building attached to it; 40 per cent of the capital expenditure undertaken by a *lespromkhoz* is usually associated with roading. It has been calculated that an increase in the skidding distance from 250 metres to 1 kilometre means that, for every 1 million cubic metres of timber skidded and extracted, instead of 170 to 200 kilometres of roads of various types being the theoretical requirement, only 40 to 90 kilometres would be needed. However, as it is the demand for the low quality, short lifespan forest tracks that is economised on, saving in capital expenditure is not necessarily in proportion to this reduction in road length. The economic benefits of the K-703 compared to the TDT-75 have been investigated by Egorov (1970) who found that, in winter, the K-703 was more economic (using the *calculated cost indicator* described in Chapter 8) with skidding distances of up to 10 kilometres and in summer over distances of up to 5.5 kilometres compared to an average skidding distance for the TDT-75 of 250 metres. If only one system was used throughout the year, the K-703 remained economical when skidding over a distance of up to 8 kilometres. Recently it has been estimated that, in winter it is always possible to use wheeled skidders because of the better ground conditions and that in summer they can be used about half of the time. If, in logging, 'only one-half of the crawler tractors are exchanged for specially constructed wheeled ones, 200 million roubles per year would be saved, 10,000 men would be released and 83,000 tonnes of metal would be saved' (Baranovski, 1977).

As well as reductions in capital costs, longer skidding distances allow a greater concentration in the upper landings. Fewer upper landings would be needed and the throughput of each could be increased. As we shall see, this type of development would suit the type of technology being employed in the loading of timber.

However, in longer tests it would seem that the K-703 does not perform very well except in the best of conditions (Borski, 1976) although the T-157 is beginning to find a place in forestry. The LT-157, a variant equipped with a grapple arm to make it into a grapple skidder in the TB-1 mould, is returning productivities per shift of about 100 to 190 cubic metres in clear felling operations (Misbko, 1980 and Tausnev, 1979). In most of these cases, however, the skidding distance appears to be quite short; in fact, the emphasis on long skidding distances seems to be lessening, which is surprising in view of the continual under-fulfilment of the plans for roading investment. Certainly, very long skidding distances measured in kilometres are very rarely referred to in the technical press.

One of the problems with the use of wheeled tractors is that, as yet, no wheeled fellers, or feller-bunchers are under development - although wheeled loaders are produced. Some argue for a four-wheel drive four-wheeled tractor for skidding and loading and a six-wheeled variant for the base for feller-bunchers and feller-skidders (Baranovski, 1977 and Varaksin and Stupnev, 1974). But one of the criteria for a system of machines is that the machines in that system should have as much in common as possible to reduce the need for stocks of different spare parts and to ease repair and servicing problems. At the moment a wheeled skidder would be at odds with this idea.

Delimbing

As was mentioned in the previous chapter, the BS-1 is slowly taking over as a specialist delimbing chain saw for use on the forest floor or at the upper or lower landing. However, special machines are being introduced to do away totally with the element of manual labour in this operation; here, again, the 1972 plan has been under-fulfilled. In 1979 35 million cubic metres of timber were delimbed using machines as opposed to the planned target of between 50 and 55 million.

Essentially, however, there are two ways of tackling the problem of the heavy requirement for manual labour in delimbing. One is to introduce machines for use at the upper landing or on the forest floor. The other is to increase the use of the whole tree extraction system and delimb the felled trees at the lower landing. Here productivity is likely to be higher as the working conditions are easier. For example, using the hand-held BS-1 delimber, productivity on the forest floor is less than 12 to 16 cubic metres per shift in most conditions. At the upper landing it might rise but only to between 15 and 20 cubic metres but at the lower landing it should reach 45 to 60 (Varaksin and Stupnev, 1974). However, in 1990 it is still planned to extract about 30 to 40 per cent of the output of *Minlesbumprom* in the form of already delimbed trunks and about 10 to 15 per cent as delimbed and cross-cut log lengths. So the mechanisation of the operation of delimbing on the forest floor or at the upper landing is of importance.

Basically there are three types of delimber now in use or undergoing tests. The first type (the LO-25) is a stationary type; this is an apparatus with a set of knife blades for delimbing through which the trees are dragged by the skidder. The delimber acts as a 'comb' in removing the branches as the trees are dragged through. Such a machine fulfils one of the criteria of Varaksin and Stupnev in that once the trees are bunched by the winch on the skidder it is the bunches which are dealt with by the delimber. But it does also mean that some of the tractor's working time is lost in the pulling of trunks through the device and skidding and delimbing require the use of the tractor in a sequential pattern of operations. However, some tests showed that labour productivity could rise by as much as one-third with the use of such a machine with productivity per skidding tractor down by only 7 per cent (Krainov *et al.*, 1972) - and this included the time spent by the tractor in clearing the brash away from the delimbing device.

The LO-25 can remove branches in spruce of up to 10 centimetres in diameter and up to 25 centimetres when delimbing broadleaved trees. The results of the LO-25 depend on the season; in winter between 70 and 90 per cent of all branches are removed whilst in summer this figure is usually down to between 50 and 70 per cent and in some cases, in spruce forests for example, it can be even lower. In this case often up to half of the branches have to be removed by hand, using axes or chain saws after this device has been used (Siritov *et al.*, 1972 and Krainov *et al.*, 1972). Against this poor performance must be balanced the fact that the LO-25 is cheap and light in weight compared to the alternatives which are based on tractor chassis and are mobile. However, this does not seem to be sufficient an advantage and this device seems to be going out of use. This is partly due to its incompatibility with grapple skidders. The use of the LO-25 involves the pulling of the trees through the device by the winch system on the skidder. As winch and choker equipped tractors go out of use, so this device becomes obsolete.

The two remaining types of delimbers that are being introduced are both mobile. This, of course, increases capital costs, but means that, first, the setting up and moving of delimbers from one site to another is independent and so less time consuming, and secondly, the engine acts as a power unit that can power the delimber's own attachments, making it independent of the skidding phase and so allowing greater flexibility and productivity from all the units in the logging process. One type of mobile delimber, the SM-2, delimbs trees individually, whilst the other type, such as the LO-39, delimbs bunches. Both types of machine use a grapple to hold individual trees or bunches and pull them through a knife system to delimb them.

Delimbing individual trees is usually more efficient than delimbing bunches but requires the picking of individual trees from the bunches delivered by the skidder. Delimbing can be troublesome if the debris in the crowns of the trees, got from their skidding over the forest floor, has a significant effect on the knives of the delimbing machine. However, such delimbers can achieve up to 98 per cent delimbing in winter conditions and up to 75 per cent in summer (Vozni, 1972). With the machines that delimb whole bunches at a time, whilst the branches on the periphery are removed by knife action, the branches towards the centre of the bunch are removed by the crushing action of the bunch as it passes through the delimbing attachment. This means that 'after delimbing, large expenditures of labour are needed to clean up the stems of the trees' (Orlov and Kochegarov, 1973).

However, if these problems can be ironed out there is little doubt that these machines could add to labour productivity. Outputs per shift of about 100 cubic metres and more are reported (Alekhin, 1976; Kanevski, 1979 and Pautkin, 1977) which means that one operator and machine would replace about six to eight manual delimbers. Of course, if some of these manual delimbers have to be kept on in order to clean up the stems of the trees after their passage through the delimber increases in labour productivity would be much reduced.

Loading at the Upper Landing

To begin with, the process of development of loaders was entrusted to *TsNIIME* which worked in collaboration with the Krasnoyarsk Factory of Forest Machinery. This factory created three loaders which were based on the crawler tractors T-100, TDT-75 and TDT-55. These are all similar looking and the P-19, based on the TDT-75, is shown in Fig. 5.2. They all operate in a different way to the typical Western type of front-end loader. In Russian they are known as *perekidnie pogruzchiki*. In Fig. 5.2 the grapple that the loader

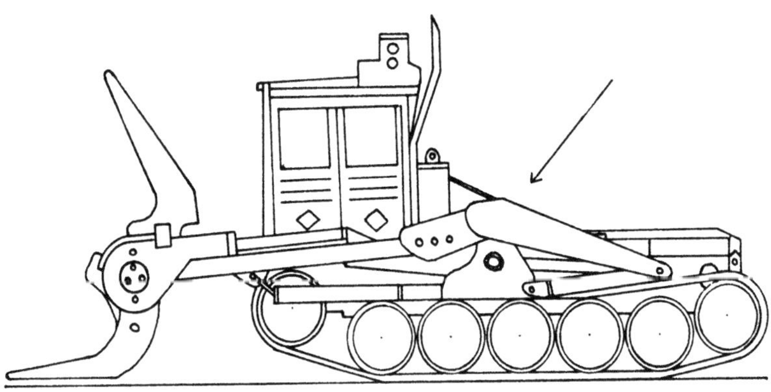

FIG. 5.2 The P-19 Loader

uses is obvious, but the action of loading is carried out by pivoting the arm about the point shown in the diagram, moving the load from the 'back' of the loader, over the cabin, to the 'front' of the machine. This is a feature which the Russians claim makes their loaders of special use in areas with poor soil characteristics. The reason is that the loader moves only backwards and forwards between the load of trees at the upper landing and the lorry, or whatever it is that is being loaded. A front-end loader would have to turn around within the confines of the upper landing, causing a greater amount of disturbance to the soil, especially if crawler tractor bases are used, and requiring a larger upper landing.

The P-2, based on the T-100 tractor, was developed in 1964, the P-19 in 1965 and in 1968 the PL-1, based on the TDT-55, was introduced for use in the northern parts of European Russia where the average timber size was less and the usual skidder was the TDT-55. By 1970, however, only 800 P-2 loaders were in use, although the P-19 was more quickly introduced, with about 1,500 in operation in that year. The productivity of these machines is quite high, with typical reports of outputs ranging in different conditions from 160 to 280 cubic metres per shift (Baranovski *et al.*, 1976). This is far higher than the output of a brigade of men working with a feller-buncher and skidder and means that these machines are increasingly being used for loading stock-piles of timber felled some time before and left at the road-side, with one loader servicing a fairly large area.

The large productivities of these machines, when compared with the output of the other elements in the logging process, has meant that the arguments for increased skidding distances and the concentration of upper landings to provide a greater throughput per landing have had more force. It is also one of the reasons for the attractiveness of 'enlarged brigades'.

The move now, experimentally at least, is to produce loaders of similar style on wheeled chassis such as the K-703 and T-157 tractors.

THE POTENTIALITIES OF THE SYSTEMS AND MACHINES

The plan of 1972 has not been met for 1980, and so its fulfilment for 1990 must be in doubt. Part of the reason for this underfulfilment is that many of the machines have not been delivered to the *lespromkhozi*. The plan for the introduction of new machines was only 53 per cent fulfilled for the period 1975 to 1980 (Solomonov, 1981). But the Soviet experts are placing a lot of hope on these new systems of machines to increase labour productivity, so the two important questions are, first, to what extent did those machines that were introduced live up to expectations and secondly, how productive will the systems be in the future when they are fully introduced?

The Performance of the Machines to Date

Nearly all of the figures for productivities of the machines discussed above are less than the planned output of the machines which were formulated when the machines were in the design phase. However, the position is somewhat worse than the figures given earlier suggest. This is because machine down-time due to breakdowns and time spent undergoing repairs is excluded from the calculations. When included it will, of course, reduce the overall productivity of the machines in question; the reliability of the machines and equipment is, therefore, of importance.

It appears that the reliability of the new machines is not very high - even when compared with the existing technology. For example, early tests showed that the LP-2 could lose between 30 and 50 per cent of work time undergoing repair (Burdin, 1973 and Emlyanov, 1972), the SM-2 delimber about 50 per cent (Svirinuk, 1971) and the VM-4 and VTM-4 up to 30 per cent (Lumenov, 1975). In one *lespromkhoz*, the VM-4 was reported to break down on average every 12 to 15 hours (Shtrek, 1978). To a certain extent this is to be expected in the introductory phase of any new form of technology, for down-time and wear-and-tear put on any machine by its operator will decrease as the machine operators become more skilled. The failure of the sawing attachment on the LP-2 and LP-18 feller-bunchers, for example, is six to eight times greater in machines whose operators have only three months' experience (Tukavin and Kopchikov, 1979). But, unfortunately, the reports in the technical press still contain figures not dissimilar to these after the machines have been in existence for up to ten years. On average, *Minlesbumprom's* tractors spend 57 days each year waiting for repair or being repaired, and loaders 106 days (Medvedev, 1977). This would represent the loss of about 200 machine shifts if a two-shift working system was being used. According to one author, out of every 100 machines in *Minlesbumprom's* enterprises, only between 58 and 60 are ready for work at any one time (Korshunov, 1977) and in 1981 Nemtsov gives figures of 51 per cent readiness for tractors and only 34 to 41 per cent for multi-operation machines such as feller-bunchers, etc.

One of the weaknesses of these machines (common to all) is the hydraulic system. About 35 to 40 per cent of all down-time may be attributed to some aspect of the hydraulic system (Goretskaya, 1978). The hydraulic pumps have a life of 600 to 700 hours. There appears to be a problem in the supply of hydraulic systems and pumps to the factories producing the machines in the first place. In one case, the director of a *lespromkhoz* ordered the changing of one pump for another in the LP-19s that he had in order to get the arm to work correctly and to increase reliability. He wrote to the newspaper *Lesnaya Promishlennost* about it (3.4.80). In reply the factory wrote:

> 'The specialists at the factory are in agreement with your opinion. One pump in place of ... (the pump in the production model) ... is of value from the point of greater reliability. However, the question of which pump to use is not decided by the Ministry of Agricultural Machinery. We are not allowed to use such pumps as you have used ... as we well know there is no spare capacity for the production of extra pumps of this type or anywhere in the production of hydraulic systems.'
>
> *Lesnaya Promishlennost*, 26.6.80

The position is further exacerbated by the general lack of spare parts for the machines, something which is common to the whole of the Soviet economy, and forestry is no exception. About one-third of all machine down-time may be attributed to the lack of instantly available spare parts (Tukavin and Kopchikov, 1979). The nine factories involved in producing spare parts for most of the forest machines give over about 7 to 8 per cent of their time for the production of spares. Calculations by some specialists have shown that this figure should be no less than 15 to 20 per cent, implying that only about half the required quantity of spare parts are being produced. Part of the reason for this shortfall lies in the nature of the planning system. To quote:

> 'At the present moment, the Ministry does not, in general, plan or control the production of spare parts as sets of parts, but gives plans to the factories expressed in physical units. With the existing goals for the machine building factories, they are not interested in the production of spares. It has come to the state of affairs that, as far as the factories are concerned, spares are a non-obligatory, second class product.'
>
> Kulikov *et al.*, 1977

Expressing plans in physical units produces the result, well known to Soviet observers, of giving an incentive to the production of the easiest-to-make article. Thus, if a certain number of spare parts are to be produced, the parts that can be most easily made will be those that will be concentrated upon, at the expense of the harder-to-make, possibly more intricate but just as necessary, ones. The authors of the quotation above argue that a system should be set up so as to provide information on the requirements for spare parts for the machines in use. This 'unreliability profile' should act as the basis for 'sets' of spares that should be produced and the plans for the production of spares should be re-specified to give incentives for the production of these sets. A similar idea for information gathering is put forward by others, who suggest that the information should pass fairly quickly to the producing factories, not having to follow a centralised path from *lespromkhoz* to *Minlesbumprom* and from there to the ministry in charge of the construction of the machines and thence to the factory in question.

Either as a result of this lack of reliability, or as a cause, the percentage of maintenance time spent at repair rather than servicing is considerable. In most *lespromkhozi* 80 to 90 per cent of the maintenance personnel's time is taken up with actual repair and only 10 to 20 per cent with routine servicing; there is also a tendency for the share of servicing to fall (Korshunov, 1977). Indeed, the decline in the relative share of servicing is even more worrying if the figures in one article are correct for the whole country that up to 15 per cent of machine down-time is the result of incorrect servicing procedures (Tukavin and Kopchikov, 1979). In fact, it is not uncommon to see authors exhorting *lespromkhozi* to service the new machines more thoroughly and to follow the factories' instructions when doing so.

Clearly, this reduction in the relative importance of routine servicing in the *lespromkhozi* is not in the long-term interests of the logging industry. However, there may be an absolute increase in the amount of time servicing, even if it declines relatively, if the amount of workers employed increases in sufficient proportion. This, it would appear, is happening in some cases, and it has the unfortunate effect that part of the good done for the overall productivity of labour by the introduction of the new machines is, in part, undone by the concomitant increase in the work-force engaged at servicing and repair. In 1976, 15,000 people above the planned number were employed at preparatory and auxiliary operations - some of which would have been repairs and servicing; this is about 7½ per cent above the planned figure (Medvedev, 1977). It also seems that the amount of work, in terms of man-days per 1,000 cubic metres of timber, taken up repairing the equipment at the lower landing has increased from 1.4 per cent of the total in 1970 to 7.4 per cent in 1975 and that the figure for repair work to the logging machines has more than doubled in the same period (Moshonkin, 1978).

The introduction of these machines means that there is a greater need to educate and train the workers in their use and a need to stabilise the work-force and prevent a rapid rate of labour turnover. Earlier the reduction in down-time of the machines operated by an experienced worker compared to a new trainee was mentioned; indeed in the hands of a new trainee these new machines would seem to become more of a liability than an asset. Not only is production down to a level less than that of cheaper equipment but also there is an increase in repair costs. Unless there is a successful result to *Minlesbumprom's* attempts to stabilise the work-force, it could find itself with machinery which, while very productive in the right hands, is less effective than the old technology when the worker does not stay long in his job.

One other problem associated with the new equipment is the level of noise and vibration. Loaders and the other equipment used on the forest floor are often criticised for having noise levels above the health norms existing. Clearly, this would be a factor that would reduce output, increase absenteeism and the rate of labour turnover - all things that the introduction of the new machines was designed to prevent.

An economic factor that will influence labour and working conditions is the effect that these machines have on costs. The use of the new systems of machines increases the cost per cubic metre. Although the number of workers is reduced and output increases, those workers who operate the machines are usually better paid than the chain saw operators and delimbers etc., whom they replace. Also the amortisation costs on the new machines are higher, and these two factors are usually enough to ensure that costs rise. In one case, costs rose from 3.68 roubles per cubic metre at the lower landing using a standard chain saw and winch and choker equipped skidder system, to 5.06 roubles with a feller-buncher and grapple skidder (Okhotski, 1976). The unreliability of the new machines adds to this problem; the use of the LP-2 in several *lespromkhozi* increased costs by up to 88 per cent over the usual technology because, in part, of the down-time of between 30 and 50 per cent. In other cases, the LP-2 'in conjunction with grapple skidders increased costs by 2 to 2½ times ... and at the same time, because of repairs and servicing work, did not achieve an appreciable increase in labour productivity' (Moshonkin, 1978). In *Krasnoyarsklesprom* the VM-4 felling machine is not productive enough to keep unit costs down and so the management suggest that a bunk and load transporting gear are grafted onto the machine to make it into a feller-skidder, in effect re-creating the VTM-4. This would increase capital costs by only 1,000 roubles and would do away with the need for the expensive LP-18 grapple skidder that has to be used with the VM-4 (Shtrek, 1978).

The result of this is that there is a tendency to move towards a multi-shift pattern of operation. Working some of the machinery for two or three shifts will increase the output but not the amortisation costs and other fixed costs. Labour costs will tend to increase in greater proportion than the hours worked because of the rates for shift work being higher than ordinary wage rates. But in spite of this, multi-shift working does reduce costs per cubic metre. In the example given above, costs were reduced from 5.06 roubles to 4.62 per cubic metre with the introduction of two shift working. Depending on the time of year, the felling machines or chain saw operators could work one or two shifts and stock-piles of felled timber would be formed; these can be loaded onto lorries by loaders working two or three shifts. The separation of the various operations that the modern equipment allows is ideal for this type of working regime. Feller-bunchers, skidders, loaders with lorries can all work independently of each other.

However, whilst this form of organisation is often mentioned as being desirable, the move to multi-shift working does not, in practice, seem to have been very marked. For example, in 1970 the 'average' logging lorry worked 1.44 shifts per day and only 1.53 in 1975 (Baranovski *et al.*, 1976). Also, the reliability of the new machines has an adverse effect on the attempt to work a multi-shift system. Results such as those shown in *Irkutslesprom* are typical. Each of a group of LP-2s, working a two to three shift system, logged 62,000 cubic metres in 10 months with an average productivity per shift of 245 cubic metres (Zverev, 1978). The productivity per shift implies a number of about 250 shifts for the ten-month period. If, on average, 2½ shifts per day should have been worked for 20 days a month, the number of shifts that should have been worked is about 500 - or the machines showed about 50 per cent down-time.

Of course, one of the effects that this move to multi-shift operations will be that the logging profession will appear less desirable in comparison to other jobs such as factory work. Forestry will, if multi-shift operation becomes commonplace, find it more difficult to attract and stabilise its work-force.

The Performance of the Machines in the Future

The problems just described will, of course, have to be remedied if the machines are to live up to their full potential. Unfortunately it is not clear that they will. There seems to be little likelihood of a change in the planning system to increase the output of spare parts in the near future and, one would be able to have more confidence in the possibility of increasing the reliability of the machines if it were not for the fact that complaints about unreliability are still seen in the technical press concerning old tractors such as the TT-4. Indeed, in the 1970s there has been a slight decrease in the actual time that these older machines are serviceable and ready for use (Baranovski *et al.*, 1976). Also, although the labour turnover has decreased slightly in recent years, it is still likely that, for the foreseeable future at least, logging will have to make do with a labour force much of which does not have long experience in the industry and possibly has no real commitment to it - see Chapter 6 for a more detailed analysis of this point.

Apart from these factors, however, other points indicate that the new machinery is likely to give rise to a slower rate of productivity increase than its technical specification suggests. First, evidence exists to suggest that the new technology (once introduced) leads only to a modest rate of productivity increase after the initial increase. In *Krasnoyarsklesprom* feller-bunchers, feller-skidders and grapple skidders were introduced. But over the ten-year period encompassing their introduction, the average annual rate of increase in labour productivity has not been very impressive except in the Predivinski experimental logging enterprise; there productivity rose by about 11 per cent per annum. Elsewhere, however, the rate of growth in the long-term has only been between 1 and 3 per cent (Kulakov, 1980), and this when, according to Varaksin and Stupnev (1974), the organisational changes and experience needed to employ fully the new machines in the most productive way should have been observed.

Another factor is to be found in the adaptability of the new equipment to the changing conditions in the Soviet logging industry. As time goes on, less and less desirable stands of timber have to be logged. Usually this is reflected in a decrease in the average tree size and a worsening of the species mix, in the sense that the proportion of coniferous species in the stand declines. As the average tree size declines so the productivity of the new machines falls. For example, as the average tree size falls from one-half to one-quarter of a cubic metre, so the length of time required by the VTM-4 to fell and bunch an individual stem falls, but only by about 25 per cent. To log the same quantity of timber, then, twice as many trees would have to be felled requiring one and a half times as long (Kapustin, 1970). The LP-2 shows a similar pattern, with productivity per shift falling from 60 to 36 cubic metres as the average tree size falls from 0.4 to 0.2 cubic metres (Perfilov, 1974a).

The second problem (the worsening of the species mix) likewise affects productivity. This problem has been investigated in the Krasnoyarsk region.

> 'In the forests of the Krasnoyarsk area, a significant percentage of the timber, from 10 to 50 per cent, consists of broadleaved species and very low quality coniferous timber, wood which has only a limited market. Therefore, in reality, logging enterprises are obliged to carry out selective cuttings - leaving the categories of timber mentioned standing.'
>
> Shtrek *et al.*, 1980

In one such area two systems of machines were tested; one consisted of one VM-4 felling machine and two LP-18 grapple skidders. The other system had two 'Ural' chain saws and two TT-4 skidders fitted with the usual winch and choker log gathering system. If all of the trees in an area were felled, the new machines were 5 per cent more productive per shift in absolute terms. However, if only 5 per cent of the trees were left standing the two systems of machines were equally productive and if 40 per cent of the timber was left, the older system of machines was about 15 per cent more productive per shift. In all three situations the labour productivity in the advanced system was higher - there were three workers employed, one with the VM-4 and two with the LP-18s, against six for the older system, two with the chain saws and two each with the TT-4s. However, the point is that the productivity of the new system was more badly affected by this type of cutting system, the reason being that the collection of bunches of stems takes less time for the TT-4 with chokers if some trees are left standing compared to a grapple skidder which must cover more of the forest to collect its load. Also the use of chain saws is easier in a situation of selective cutting when some of the trees are left standing than is the use of the more cumbersome felling machine.

OVERALL INCREASES IN LABOUR PRODUCTIVITY

The emphasis in the Soviet Union seems to be on increases in labour productivity that the logging systems described can bring about. Of course, logging (getting the timber from stump to upper landing) is not the whole of the story. On its own the plan for the period 1972 to 1990 (*Minlesprom*, 1972) would have increased labour productivity in the whole of the logging industry by only 2 per cent per annum from 1972 to 1980 and by only one-half of a per cent from 1980 to 1990 if no increases in other operations were recorded. Clearly the overall productivity increase would depend upon the changes in labour productivity in the other phases of the logging process - extraction to the lower landing, lower landing work and the various auxiliary and preparatory operations (cf. Table 4.2 in the previous chapter). As has already been mentioned when the 1972 plan was described, if productivity increases in line with those envisaged in the plan for logging were reproduced in all the operations, the productivity of labour would rise by 11 per cent per annum from 1972 to 1980, and by 6 per cent per annum from 1980 to 1990. But if the felling/ skidding/loading processes are not maintaining their projected growth rates in terms of productivity, what of the other operations?

Transport to the Lower Landing

The 1972 plan contained elements for the future of transport to the lower landing. In Table 4.2 in the previous chapter, it was seen that transport accounted for only about 5 to 6 per cent of the total labour input in logging. Table 5.1 shows that there is very little unmechanised labour input into the extraction process and so there is little scope for immediate productivity increases as there is in skidding where the use of grapple skidders can double productivity. The plan contains only the projection of increased power of the lorries used in the operation and an increase in the use of lorries as opposed to narrow-gauge railways. New KAMAZ and KrAZ tractor units have been introduced which allow a greater load per trip and higher speeds.

However, a major factor here is the nature of the extraction process; the extraction distance is tending to increase each year at a rate of between 0.5 and 2 kilometres which will be accentuated by any increase in concentration in lower landings, as was discussed in Chapter 3. Also, the longer the extraction distance, the more likely the extraction process is to be troubled by the weather. The programme to build all-weather roads will, to some extent, reduce the seasonal variation in output from *Minlesbumprom's* forests and so help increase productivity. However, in the short-term, an increase in extraction distance of 1 kilometre decreases the productivity of logging lorries extracting over a distance of 40 to 50 kilometres by between 1.5 to 2 per cent (Gorbachevski, 1973). As one lorry is associated with one driver, this is equivalent to a 1.5 to 2 per cent fall in the productivity of labour. Some estimates predict that, by the year 2000, average extraction distances will be in excess of 100 kilometres compared to the present figure of about 40 (Tatarinov, 1978). This would effectively halve the productivity of extraction if no change in technology of the process occurs.

To overcome this, all-weather roads would have to be built, but the road building programme is almost always referred to in the technical press as not having been fulfilled. It is either noted as being insufficient to allow the full exploitation of the winter period (*Lesnaya Promishlennost*, 2.8.79) or else as being insufficient to allow summer and autumn extraction (Tsibayev, 1980). So it would seem that large increases in the productivity of labour in extraction in the future are unlikely. It is more likely that any increases in power and speed of lorries, and improvements in roading are going to be neutralised by increases in the extraction distance, leaving productivity relatively constant.

Lower Landing Work

A section on lower landing operations is contained in the 1972 plan of mechanisation. Table 5.1 shows that operations at the lower landing contain a great deal of manual labour - especially loading and sorting. As was mentioned in the previous chapter, work at lower landings is harder to generalise about than work on the forest floor. The plan for the development of lower landings hinges around two main points. First, there is to be an investment programme in equipment to do away with the elements of manual labour and, secondly, there is to be a concentration of production by closing smaller landings. Larger landings are more efficient and have a sufficient throughput to make the investment in the capital equipment needed to do away with manual labour economic. In smaller lower landings, with annual throughputs of less than 100,000 cubic metres, it is uneconomic to use anything except electric hand saws for delimbing and cross-cutting and low levels of mechanisation for loading and sorting. The optimum size of a lower landing is often mentioned as being between 300 and 600 thousand cubic metres per annum (Varaksin and Stupnev, 1974), a size which would reduce costs to a minimum. However, this size minimises only the costs of processing within the lower landing itself, and not those of the whole logging process. The effect of greater concentration has been mentioned above in the section on extraction.

The increase in capital equipment that is to be carried on concurrently with the concentration in lower landings involves the introduction of semi-automatic saw lines. The smaller lower landings are to have lines that cross-cut individual logs into required lengths by using single saws past which the logs pass one by one, the larger landings are to use lines having banks of circular saws to cut logs in one go into the lengths required. Wheeled loaders are likely to find a wider use in the better conditions of the lower landing than they do at the upper landings.

The developments in lower landings tend not to be discussed in such depth and with the same generality as are those involving logging. However, some generalisations are possible. First, the problems of reliability appear to be similar to those experienced by the users of the new logging equipment on the forest floor; this seems to apply especially to the new saw lines for which spare parts are difficult to obtain. Secondly, a factor that Stupnev (1978b). points out is important. He states that the basic use of lower landings is to provide a sorting stage in the logging process so that different species and sizes can be sorted and the correct ones delivered to the final user. In the Soviet Union, he says:

> 'logging ... is, as a rule, in natural forests with an exceptional mixture of sizes, species and quality characteristics.'

This will tend to increase the necessity of sorting operations and the creation of stock-piles - which are heavily labour intensive. Also, the variety of species, and therefore tree forms and shapes, will tend to reduce the possibility of mechanisation to the full. Thirdly, as more delimbing is transferred to the lower landing from the forest, the amount of work that has to be carried out there will, of course, increase.

So it would seem likely that the mechanisation of lower landing work will be at least as difficult as is the mechanisation of work on the forest floor. Performance to date is a little more difficult to assess. In the past few years the increase in the productivity of labour at lower landings has been slower than that experienced in the felling, skidding and extraction operations. Some figures suggest that, between 1969 and 1974 the productivity of labour actually declined by about 10 per cent (Baranovski *et al.*, 1976). The 1972 plan was supposed to increase labour productivity at the lower landing by 1½ times between 1972 and 1980 (or about 5 per cent per annum) and by 2 times between 1980 and 1990 (about 7 per cent per annum). This plan required that about 1,000 semi-automatic saw lines should have been in operation by 1980; in fact, the number seems to have been nearer 750 (Anon, 1979). Also, the plan required that between 40 and 50 per cent of the output of *Minlesbumprom's* lower landings should have been produced by such lines. In 1979, the figure was only 20 per cent and it would seem, therefore, from these figures that the investment programme is about as far behind schedule as is the plan for work on the forest floor.

Preparatory and Auxiliary Work

Looking at the figures in Table 4.2, preparatory and auxiliary work accounts for about half of the labour input into the whole logging process. The breakdown of this category of work into more specific operations is given in that table for *Archangelsklesprom* for 1963. Apart from figures for repairs, it is unlikely that the overall percentage composition of preparatory and auxiliary work has greatly altered. In 1963 the figures for the individual elements of this phase of the work were: preparation of the felling area, 1.7 per cent of the total labour input into logging, road maintenance, 8.4 per cent; building of logging roads, 4.3 per cent; maintenance and repair of equipment, 20.9 per cent; the transport of workers to the work site, 1.8 per cent; and other categories of work, 9.2 per cent.

Preparation of the felling area. Whilst this operation accounts for a small proportion of the total logging effort, it was hoped that the introduction of mechanisation would reduce the need for this type of work. For example, the need to fell dangerous trees before general felling operations could begin, it was hoped, would be avoided by the use of machines with enclosed cabs. Some of the work in this category, delimiting the felling area, surveying the area to discover the average tree size and standing volume for planning purposes etc., is going to be unaffected by the introduction of the new machines. Also, in some cases, large tree sizes cannot be felled by the felling attachments on the new machines, such as the LP-2 and LP-19. This means, that, unless they are felled by a crew using the conventional types of chain saw, valuable timber would be left standing. This would tend to compensate for any reduction in the amount of work needed to fell dangerous trees before felling operations began.
It seems unlikely, therefore, that any great change in the work required for preparing the felling area is going to occur.

Road maintenance and building. Taking these two categories of work together, the points made earlier would seem to rule out any great reduction in this area in the number of man-days needed to maintain and build roads. Indeed, the calls in the Soviet press to increase the amount of work done in roading suggest that it will increase. The possibility of a radical increase in the skidding distance, leading to a reduction in roading requirements seems to have disappeared from the alternatives being considered. Any increase in the roading programme will also bring about a reduction in output from many of the conventional logging machines for many of them are also used in roading. Already, 6 per cent of the working time of *Minlesbumprom's* skidders is taken up in road building and maintenance, and loaders are often used for this purpose as well (Baranovski *et al.*, 1976).

Maintenance and repair of equipment. The worrying increase in repair and maintenance work and the personnel needed to carry it out has been outlined in an earlier section of this chapter. It is likely that, as time goes on and the equipment becomes more reliable, this requirement will tend to decline. But, intrinsically, it seems probable that the new systems of machines will need more associated repair and maintenance work and personnel. The replacement of, say, two chain saws, a tractor with winch, a few axes for delimbing and a couple of A-frames and winches for loading by a system of machines made up of a feller-buncher, grapple skidder, mobile delimber and tractor based loader means that the amount of servicing needed is almost certain to increase, even if the new equipment turns out to be as reliable as the equipment it replaces.

It would seem most unlikely that any reduction in the number of man-days needed for this category of work can be expected in the future. Indeed, this particular element of preparatory and auxiliary work is most likely to increase both relatively (as a percentage of the total work needed in the logging process) and in absolute terms.

The transport of workers. As most workers are based at or around the lower landing and not the upper landing, where the area being logged is situated, the increase in the extraction distance that is occurring in Soviet forestry will lead to an increase in the time required to transport workers to their work. Any increase in the speed of transport used is likely to be offset by the increasing distance and so it is unreasonable to expect any decline in the number of man-days consumed in this activity. The use of the *vakhtovi* method does not seem to be increasing sufficiently to affect the figures referring to this operation.

Other work. Of the types of operations included in the 'other work' category, one activity is worth noting. In recent years, there has been an increase in the number of workers employed in calculating the actual output of the various workers in a logging brigade for the purposes of calculating bonus payments (Shcherbakov *et al.*, 1978). As the various elements of the brigades in the logging process become less inter-dependent, so the need to measure individuals' outputs instead of the output of a group increases. This will cause an increase in the amount of work in this particular category. Some authors argue for a change to the use of the weight of timber as a measure of output for bonus payment purposes and not volume, which is a difficult and time-consuming thing to measure.

CONCLUSIONS AND IMPLICATIONS

The introduction of any new equipment can bring about an increase in productivity in two ways. First, it can be directly more productive than the equipment it replaces and, secondly, it can allow a more productive organisation to be employed. In Chapter 4 it was seen that, whilst some Soviet experts claimed that the latter was an important factor in the growth of labour productivity in the post-war period, a detailed analysis tended to uphold the view that productivity increases were closely linked to the nature of the capital equipment itself. If this is a true representation of the evidence presented it is reasonable to assume that the pattern of productivity growth in the future will be influenced by much the same sort of factors. That is, an estimate of future productivity changes can be had by examining the nature of the machines being introduced. This has been undertaken in this chapter, and we are now in a position to suggest possible future changes in productivity in Soviet logging. This is an important projection to make for, coupled with the predictions of the size of the work-force available to be made in the next chapter, an estimate may be arrived at to show the potential that the Soviet forest industries have for future expansion, if any, and this affects not only the internal economy, but also their export potential.

Two extremes can be defined to the range of possible paths that the productivity of labour could take through time. The first, an optimistic one, is that the production of the required machines speeds up to the rate foreseen in the 1972 plan (*Minlesprom*, 1972), the reliability of the equipment improves as its 'teething problems' are worked out and the labour force is stabilised so that the turnover of workers is reduced, giving the forest industries a skilled work-force that is able to extract the maximum output from machines whilst causing a minimum of wear-and-tear. In this case, it would be reasonable to assume that the productivity levels reached would be those envisaged in the 1972 plan. An alternative path, a pessimistic one, is that the introduction of machines remains at its present slow rate, their intrinsic reliability remains unaltered but, because the level of labour turnover fails to alter, the inexperienced and 'itinerant' work-force causes the productivity levels of the new machines, in many cases, to be no higher than the equipment it replaces.

These two possibilities are analysed in Table 5.9. The operations in the logging process are laid out as they were in Table 4.2 and in the first column the figures for the number of man-days needed to produce 1,000 cubic metres of timber in the Archangelsk region in 1970 are reproduced; the figures for the various constituent elements in preparatory and auxiliary operations are calculated on the assumption that they constituted the same proportion of the total in 1970 that they did in 1963. In the second column are figures that compensate for the fact that the productivity of labour in *Archangelsklesprom* is about 7 per cent less than the average for the Soviet Union as a whole. The figures in column two might, then, be thought of as being Ministry averages for 1970.

The two other columns are derived from column two. Column three represents the position projected for 1990 under the optimistic assumptions. Here it is assumed that the 1972 plan is fully implemented by 1990 and that the productivity levels mentioned in the plan are fulfilled. Productivity in felling and skidding is assumed to be 35 cubic metres per day (a figure derived earlier in this chapter) which gives rise to a requirement of 28.6 man-days to fell and skid 1,000 cubic metres. The 1972 plan contained no projections for productivity in extraction, and so the figure in column three is an extrapolation based on the assumption that the productivity of labour continues to grow at the same rate between 1970 and 1990 as it did between 1963 and 1970. The 34.2 man-days of labour required for work at the lower landing represents a growth in the productivity of labour over the 1970 figure of 5 per cent per annum until 1980 and 7 per cent per annum between 1980 and 1990, the projections in the 1972 plan. This gives rise to a total number of man-days needed for basic operations of 73.6. In the preparatory and auxiliary operations, projections are a little more difficult. It is assumed that the number of man-days required for preparation of the felling area falls slightly to 6.3, representing the fact that felling of dangerous trees, etc., is no longer needed with the

TABLE 5.9 Projections of Productivity in the Logging Industry

	1970 (1)	1970 (2)	1990 (3)	1990 (4)
BASIC OPERATIONS	305.0	285.0	73.6	148.2
felling/skidding	158.8	148.4	28.6	49.5
extraction	28.9	27.0	10.8	27.0
work at lower landing	117.3	109.6	34.2	71.7
PREPARATORY AND AUXILIARY OPERATIONS	243.2	227.4	202.5	235.1
preparation of the felling area	8.9	8.3	6.3	8.3
road maintenance	44.3	41.4	33.9	41.4
building of logging roads	22.6	21.1	17.3	21.1
maintenance and repair of equipment	109.8	102.6	100.0	110.3
transport of workers	9.5	8.9	8.0	6.9
other	48.3	45.1	37.0	45.1
TOTAL	548.3	512.4	276.1	383.3

Column (1) refers to the number of man-days required to prepare 1,000 cubic metres of timber ready for delivery from a lower landing in *Archangelsklesprom* and is taken from Table 4.2 in the previous chapter. For the significance of the other figures see the text.

use of the new machinery. The figures for road maintenance and road building, 33.9 and 17.3 man-days respectively, are prepared on the assumption that the productivity of labour in these operations grows at a nominal 1 per cent per annum. The figures of 100 and 8 man-days per 1,000 cubic metres for repair and maintenance of equipment and the transport of workers represent slight improvements over the 1970 figures. It is unlikely that radical changes in the productivity of labour in these areas are going to occur, even under favourable assumptions, for reasons outlined above. The figure of 37.0 man-days for other operations incorporates a nominal 1 per cent growth rate over the 1970 figures. Together these operations give rise to a requirement of 202.5 man-days per 1,000 cubic metres of timber logged.

Overall, then, under these optimistic assumptions, 276.1 man-days would be needed to fell, skid, extract and process 1,000 cubic metres at the lower landing in 1990. This compares with the figure for *Minlesbumprom* as a whole in the early 1970s of 512.4 and represents an overall growth rate in the productivity of labour of 3.1 per cent per annum. By far the major part of this growth occurs in basic operations, where an average annual growth rate of 7.0 per cent is registered between 1970 and 1990. However, this projection does seem to be a very optimistic one and is possibly better interpreted as a maximum growth rate of labour productivity than the logging industry in the Soviet Union could achieve. Already, in 1980, the 1972 plan is more than five years behind schedule and there seems to have been little, if any, growth in productivity in the preparatory and auxiliary operations and also, apparently, in basic operations (Solomonov, 1981).

Turning now to the pessimistic forecast, column four contains the figures derived from column two that are based on pessimistic assumptions. The figure of 49.5 man-days for felling and skidding is based on the assumption that the introduction of technology continues at the present rate; earlier it was shown that already the 1972 plan is about five years behind schedule. Assuming that it is completed in the year 2000 rather than 1990, the productivity of labour in each year to 1990 will only grow at about 5.6 per cent The figure in Table 5.9 reflects this. The figure for the number of man-days needed for extraction is based on the assumption that the extraction distance increases and exactly compensates any increases in the power and capacity of the means of extraction. The figure for work at lower landings, 71.7 man-days per 1,000 cubic metres, is based on the assumption that the growth in the productivity of labour in this area is about 3 to 4 per cent less than that in felling and skidding. This seems to have been the trend over the last few years. This leads to a figure for basic operations of 148.2 man-days per 1,000 cubic metres compared with 73.6 for the optimistic assumptions.

The difference between the two sets of assumptions in the preparatory and auxiliary operations is less marked. In the preparation of the felling area it is assumed that the productivity of labour remains unaltered reflecting the view that what is saved in not having to fell dangerous trees before an area is logged over is exactly matched by having gangs of men with chain saws to fell trees which the advanced felling machines are unable to handle. The figures for road maintenance, road building, transport of workers, and other operations in this category are the same as those for the Ministry average in 1970. Here the assumption is that any increases in the productivity of labour because of the introduction of new equipment are not made apparent because of the increases in the absolute amount of work in these categories that has to be done. Road building and maintenance are likely to increase in the future and the figure for repair and maintenance of equipment of 110.3 man-days is based on figures quoted earlier that 7½ per cent more personnel were employed in this branch of operations than was planned. This gives a total figure for preparatory and auxiliary operations of 235.1 man-days per 1,000 cubic metres.

Overall, then, the number of man-days needed to produce 1,000 cubic metres of timber under the pessimistic assumptions is 383.3 compared to 276.1 under the optimistic assumptions. These figures represent a growth rate in the productivity of labour of only 3.3 per cent per annum in basic operations and an overall growth rate of only 1.5 per cent.

Some recently published figures might help to suggest which of these two forecasts is the more likely. Research was carried out at a *lespromkhoz* in *Tumenlesprom* (Orlov and Skomorokhova, 1981). The *lespromkhoz* in question is "a repeated leader in the industry in the innovation of progressive forms of technology and methods of labour organisation". However, despite its good record and the fact that its labour productivity has been "consistently greater

TABLE 5.10 Productivity of Labour in *Tumenlesprom*

	1976	1980
BASIC OPERATIONS	125.3	123.6
felling/skidding	44.6	33.9
extraction	19.8	22.8
work at lower landing	60.9	66.3
PREPARATORY AND AUXILIARY OPERATIONS	153.5	226.4
TOTAL	278.8	350.0

Source: Orlov and Skomorokova, 1981.

than the existing norms in the industry", productivity decreased over the tenth five-year plan. The figures are shown in Table 5.10. Again, they are in the form of the number of man-days required for the production of 1,000 cubic metres of timber at the lower landing.

Looking at these figures, it is apparent that, although the *lespromkhoz* is efficient compared to others in the country, it is becoming less efficient at extraction and at work at the lower landing. Also, although it seems to be using the new logging technology efficiently in felling and skidding, there has been a dramatic increase in preparatory and auxiliary operations. This has been mainly due to a large increase in repair and maintenance work.

Remembering that these figures represent an efficient *lespromkhoz* it would appear that of the two forecasts in Table 5.9 the pessimistic one is the more likely. Whilst the machines being introduced are, in themselves, often sufficiently productive to fulfil the figures in column three of the table the problems of reliability are likely to continue at least into 1990, and this will reduce the effective productivity of the machines to give the figures shown in column four, or at least figures close to them. So if the two sets of figures which are presented for 1990 are taken as being maximum and minimum levels of future performance, it is likely that productivity growth through time in the logging industry will be closer to the pessimistic forecast than the optimistic one.

This means that labour productivity is likely to grow by between 1.5 and 3.1 per cent per annum. This compares with the present planned level of growth of 3.2 per cent, and probably the actual growth rate will be towards the bottom of this range, implying that the eleventh five-year plan will not be fulfilled in this respect.

An Aside on Western Technology

Experts in the Soviet Union have carried out a few tests to compare their equipment with Western models. One feature that the Western technology often possesses is a capacity to carry out more operations than its Soviet counterpart. In the West the short wood system is popular and feller-delimber-bunchers, delimber-crosscutters-forwarders are common machines. This point is recognised in the Soviet Union, but experts there point out that feller-delimber-bunchers and other such machines would have to be able to operate with both coniferous and broadleaved species. This imposes difficulties on design that are avoided in the Western experience (the Russians claim) where a more homogeneous species composition is found (Nemtsov *et al.*, 1979).

Although the Soviet experts appear to want to move as many operations off the forest floor in order to improve working conditions, feller-delimber-bunchers would be no worse for the operator to use than the feller-bunchers or feller-skidders being introduced now. But feller-delimber-bunchers are only just in the development stage in the Soviet Union. Possibly one of the reasons for this is that the attachments on feller-bunchers and feller-skidders now in use have been sufficiently unreliable to discourage the development of more complex apparatus.

Certainly, as a little under one-half of the timber to be felled is to be delimbed before it reaches the lower landing, there would seem to be a need for such a machine and it would certainly be cheaper than having to introduce separate feller-bunchers and delimbers.

Authors point out the higher reliability of Western equipment when compared to Soviet machines. For example, in comparative tests between a Drott feller-buncher and the LP-19, it was reported that they had nearly identical productivities on an hourly basis, but that the Drott was more reliable and so gave a greater output over a long period. Also, a point worth noting, the LP-19 did less damage to the base of the trees it felled than the Drott because of its chain saw based felling head. The Drott uses a knife/shear arrangement. As this is a common system on Western machines, it is likely that the Soviet machines, which use chain saws in nearly all of their configurations, have the edge over the conventional Western models in this respect (Sakharov and Sokikas, 1980).

Another point that Soviet authors mention with respect to Western technology is the high driver comfort of most of the models. The convenience of operation, the layout of the controls, the protective features of the construction of the cabin, the low noise levels and the air conditioning often found is described as being of a very high standard, and, by implication, better than the corresponding features on Soviet equipment (Korobov and Ribakov, 1977 and Shcherbakov, 1978).

Apart from these features, it is usually noted that in general Western machines are based on wheeled bases with higher powered engines than Soviet models - often 200 horse-power or more. The engine power and wheeled bases give the machines a higher speed than the Soviet ones and so either increase productivity or allow an increase in the skidding distance to reduce roading costs as was described earlier. The low power of the Soviet machines was noted in the previous chapter. Part of the reason for the continued low power lies in the fact that the under-powered tractors serving as bases for the new machines were seen as a quick way to mechanise the logging process, rather than first developing a higher-powered tractor unit. But there are no higher-powered units coming into use in any large amount, only the K-703 could fit the bill and this has not been introduced in large numbers.

Finally, a system where felling, delimbing, cross-cutting, bunching and forwarding are performed by one machine, often known as a 'tree harvester', is a system sometimes found in advanced logging operations in the West. But again, maybe because of reliability, it does not yet find use in the Soviet Union, despite its projection as system L8 in the 1972 plan. Although the author has seen no reports of such a machine in recent years, Perfilov (1974a) does report on a LK-4 tree harvester being designed by *TsNIIME* and based on the VTM-4 and tested in *TsNIIME's* own experimental *lespromkhoz*. It has a productivity of between 40 and 50 cubic metres per shift but these reports were dated 1972 and it is not clear whether these figures were planned productivities or results of actual tests.

Appendix to Chapter 5

If for the ith logging system

p_i = the productivity of labour in cubic metres per shift

s_i = the number of shifts per annum

w_i = the number of workers working in the system

v_i = the percentage of the total volume logged to be accounted for by the ith system

and if for the logging industry as a whole

V = the total volume logged in a year

W = the total number of workers in logging

P = the overall productivity of labour in cubic metres per shift

then

$$p_i w_i s_i = Vv_i/100$$

by re-arranging

$$w_i = Vv_i/100p_i s_i$$

by summing over all systems

$$W = \Sigma Vv_i/100p_i s_i = V/100\Sigma v_i/p_i s_i$$

if s_i is constant for all systems = s

$$V/Ws = 100/\Sigma v_i/p_i$$

but for the whole of the logging industry

$$P = V/Ws$$

therefore

$$P = 100/\Sigma v_i/p_i$$

6
The Labour Force in Soviet Industry

INTRODUCTION

The forest industries in the Soviet Union are an important employer of labour as about 3 million people are employed in the forest industries as a whole (Glotov *et al.*, 1975). However, it is in this area that many of the future problems that Soviet forestry will have to face are to be found. Western analysts are pointing to the fact that in the future the increase in the work-force will be less than that experienced in previous years. In the forest industries this is going to manifest itself as an actual decline in the work-force. Already this reduction in the supply of labour has had its effect on the direction of research and development into logging equipment. In Chapters 4 and 5 it was seen that an emphasis was placed and still is being placed on increasing the productivity of labour, even if costs rise as a result.

It is possible that the mechanisation of operations in logging and timber hauling and conversion are only going to be effective as a 'holding operation'. It is likely that increases in the productivity of labour brought about by new investment will only compensate for reductions in the supply of labour to logging. This means that it is probable that in the future the output of timber from Soviet forests will not increase to any great extent. Thus, if the analysis to be presented in this chapter is correct, any increase in the output of timber products in the Soviet Union must come from an increase in the fraction of the harvested trees actually converted into some sort of final product.

As well as looking at the potential supply of labour this chapter will also consider other topics of interest such as the use of penal labour in logging, the use of women in the forest industries, elements of Soviet labour law and the incidence of industrial injury in logging.

A BRIEF HISTORY

To talk of a specific labour force for the forest industries in pre-revolutionary Russia is possibly misleading as agricultural workers would often work in winter at logging specific areas for mill owners. However, it makes much more sense to talk about a labour force in the sawmilling industry, which, in Tsarist times, existed mainly to serve the export trade and, to a lesser extent, the industrial development of the country. By 1912 the number of workers engaged in the sawmilling industry was 117,000, employed at just over 2,000 enterprises (Algvere, 1966).

After the revolution and civil war, the position was probably not radically different in terms of the numbers working in forestry. However, as time progressed the forest industries began to become organised and acquire a permanent work-force. By the end of the 1930s over 1,300 thousand were employed in the Commissariat of Forestry, and this Commissariat by no means accounted for all timber output of the Soviet Union (Jasny, 1959).

Penal Labour in the 1930s

There is much evidence that both criminals and political prisoners were used in logging camps. In 1931 H.R. Knickerbocker published a book which described his trip to the Soviet Union, and describes the use of *kulaks*, the middle-class peasantry, in the Volga-Caspian Timber Combine. They were employed to fell and transport timber, not being allowed in the sawmills or near other machinery for fear of sabotage. They were there as part of the process of 'dekulakisation' - an attempt to rid them of their bourgeois attitudes as they had opposed collectivisation of agriculture. The President of the Combine is quoted as saying;

> 'They are not convict labour. These fellows are only sent up here to learn to adapt themselves to the Soviet system. The only difference between them and the other labour is that they have got to stay here.'

During the late 1920s and 1930s governmental pressure was exerted on managerial bodies of the collective farms to fulfil voluntary contracts to supply labour to industry and one of the industries which it was hoped would benefit from this policy was logging. However, for the forest industries at least, this policy does not appear to have been successful. Swianiewicz (1965) states that for the first five-year plan the Government had tried to mobilise much of the labour force needed for logging in the remote areas of the country on the basis of these contracts with the collective farms. However, he reports that in 1930 only 50,000 men had been mobilised instead of the 900,000 needed.

Part of this shortfall was made up by assigning some of the planned timber output to the security organs. The OGPU, and later the NKVD, used some of their inmates to fell timber. Ex-prisoners' reports on this are not uncommon, for example Yevgenia Ginzburg in her memoirs (1968) devotes a chapter to her experiences felling timber in the Far East of Russia. However, it should not be imagined that arrests were carried out specifically to provide a labour force for lumbering. As Conquest (1971) points out;

> '... engineers, scientists, doctors and lawyers were not arrested simply to provide a corps of incompetent lumberjacks.'

To many, being sent to a lumbering camp or being included in a lumbering gang was very similar to a death sentence. The food ration was small and of poor quality and the work hard and usually carried out in extreme conditions. Any under-fulfilment was usually answered with a decrease in the person's rations. Table 6.1 shows the size of ration for camp inmates in areas of extreme climate. It is reported that inmates actually engaged in tree felling received a slightly greater ration (Conquest, 1971) but as Solzhenitsyn (1963) points out, often prisoners did not receive the full ration.

TABLE 6.1 Rations for Prisoners Engaged at Hard Labour

100 per cent of the norm	800 grammes of black bread, plus soup and buckwheat
non-fulfilment of the norm	400 grammes of black bread plus soup
disciplinary ration	300 grammes of black bread

Source: Conquest, 1971

Possibly the fear of being sent to lumbering camps, especially in the Far East, has led former prisoners to overestimate the role of felling and timber hauling in the productive activities of these camps. Conquest (1971) assesses that fewer prisoners were engaged in lumbering in the early 1940s than in mining or general construction, for example. Possibly the best information about the use of prison labour in the forest industries comes from Jasny (1959) who presents figures purporting to come from the plan for 1941, which was not put into effect because of the war. According to these figures, the NKVD was to be responsible for the felling of 34.7 million cubic metres of timber and firewood, the transporting of 37 million cubic metres of timber from the forest, the production of 2.8 million cubic metres of sawn timber and the production of 189.3 million roubles' worth of furniture at the prices ruling at the time.

Jasny calculates that, if the productivity of labour was the same for the NKVD and the Commissariat for Forestry, the NKVD would have needed 260,000 people to fulfil the plan, 240,000 working at felling, transporting and related jobs and 20,000 as factory labour. The assumption that prison labour was as productive as other labour has not been accepted by other authors. Ginzburg (1968) tells that the norm for herself and her helper was 8 cubic metres per day to be felled. She reports that, although some women could fulfil their norm, mostly peasant women used to manual labour, she often only managed 20 per cent of this norm and as a result her food ration was reduced.

Swianiewicz (1965) suggests that the number of prisoners working in the forest industries in 1940 was 400,000. He reaches this conclusion on the basis of Jasny's estimate of 260,000. Whilst adjusting the figure for factory labour, his main additions to the number come in the form of an extra one-third to account for the reduced productivity of labour and an extra one-sixth to account for the number of prisoners who died during the year. This gives him an estimate, when rounded, of 400,000 prisoners or about 6 per cent of the estimated camp population of 1940 of 6.5 million.

However, there is the possibility that even this figure is too small. Swianiewicz agrees with Jasny that the technology of production in the camps was the same as that outside and so the only reason for reducing the figure for the productivity of labour was the nutritional and motivational one. This conclusion is based on figures for the amount of mechanised transportation to be undertaken by the Commissariat of Forestry and by the NKVD. However, transportation is different to skidding. Assuming that the overall organisation of logging in NKVD enterprises was the same as that used in ordinary concerns, i.e. timber was skidded to an upper landing of some sort and then transported to a lower landing or direct to a final user, the level of mechanisation of skidding is of importance in determining the overall level of productivity. In ordinary logging concerns horses were mainly used for skidding and in 1935 only 9 per cent of timber was skidded by tractor. But in labour camps the position is not so clear. As a result of his researches, Conquest states;

> 'In the labour camps there were seldom tractors or horses, and sleds of wood were pulled by the prisoners. If the team was made up of men, five were harnessed, if of women, seven. A Polish journalist who served in the Pechora camp reports seeing several hundred women carrying heavy logs ...'

In contrast, however, Ginzburg states that tractors arrived one day to skid out the timber that they had felled which they had been allowed to leave on the forest floor.

Thus, the actual level of mechanisation used in skidding is a matter of doubt, but it would seem that the operations in the ordinary concerns were less labour intensive than in the camps. If all of the timber skidded by NKVD camps used human-powered sleds, then the estimate of 400,000 workers engaged in logging would be too small by a factor of one-third. This figure is based on the assumption of a ten-man gang producing 5 cubic metres of felled timber at the lower landing - this includes fellers, delimbers, loaders, and horse handlers. Assuming a 300-day working year, this gives an annual output per gang of 1,500 cubic metres, and would require about 23,000 such gangs to produce the needed output of 34.7 million cubic metres, or about the number of workers that Jasny estimates would be required. If a five-man sled hauling team has about the same productivity as one man and a horse the lower level of 'mechanisation' in the labour camps would require fourteen men to a gang, as opposed to ten in the normal set of circumstances. Applying this corrective factor to Swianiewicz's estimate which itself already contains an allowance for reduced productivity gives a figure for the number of workers needed in NKVD logging camps to fell and transport 34.7 million cubic metres of timber of 560,000. As this is a figure that assumes that all the timber was skidded by muscle power, it seems safe to conclude that the actual number of prisoners employed in logging in 1940 was between Swianiewicz's estimate and my own, that is somewhere between 400,000 and 560,000.

Conquest's book contains much anecdotal evidence of life in

logging camps; he reports that some ex-prisoners state that the death rate in logging brigades could be as high as 50 per cent, although 30 per cent per annum was apparently nearer to the norm. In order to maintain the output from the prisoners, the food ration was made dependent upon the overall output of a gang or brigade of workers. Therefore any underfulfilment of the norm had an effect on every member of the group. This imposed some sort of self-discipline on the prisoners to work harder and not to try to 'beat' the system. The work norms seem to have been very high in the circumstances. Ginzburg, as had already been noted, was presented with a norm of 8 cubic metres per day to fell which she found very difficult to meet; indeed, in some logging camps it is reported that work continued until midnight to fulfil the norm. This was made even more serious by the fact that the ration depended on the output of a prisoner; inability to fulfil the norm would lead to a reduction in the ration and so a weakening of the prisoner making it impossible for him or her to fulfil the norm in the future. To quote at length from Conquest;

> 'Even comparatively mild lumber camps were great killers. Finnish prisoners who were experts in the matter held that the norms prevailing were impossible even for the best fed workers. They could only be brought down even to a barely tolerable level by various forms of cheating and bribery - using the same logs over again, by sawing off the stamped end, and so forth. [One inmate] never came across a prisoner who had worked in the forest for more than two years. After a year they were usually incurable, and following transfer as 'goners' to lighter work were retired to the morgue.'

The operations of the NKVD were not evenly distributed around the Soviet Union; in 1941 the NKVD was to account for about 12 per cent of the total timber production. In central European Russia the NKVD was not active at all in lumbering, whilst in the Leningrad region only 5 per cent of output was to come from camps run by the security organs. In the area of Archangelsk, however, slightly over one-quarter of all timber was to be produced by prison labour, and in the Komi ASSR the figure was over a half. In Siberian Russia and in the Far East the percentage was probably even greater than that.

As well as actual lumbering activities, prisoners were also involved in the construction of camps and factories in the more remote areas of the country for the use of free labour. Conquest estimates that 3.5 million prisoners were involved in general construction in 1941. Solzhenitsyn, in his novel 'Ivan Denisovich', describes prisoners marching past a 'wood processing factory ... the workers' settlement ... the new club ... convict built in its entirety.' If this is the case the contribution of penal labour to the forest industries is even greater than the not inconsiderable proportion of timber logged by prisoners.

In the immediate post-war years, the camp population was swelled, possibly by as much as 15 million, by a large influx of ex-servicemen, civilians from occupied areas, certain Soviet nationalities and German prisoners-of-war. However, these prisoners seemed to have had a high mortality rate and so the actual numbers in the camps would be somewhat less than the figures suggested by the rate of influx. So, in the late 1940s it is probable that the camp population played an important role in the forest industries. By the early 1950s, however, the supply of camp inmates probably began to dry up, and with the death of Stalin and the relative liberalisation, the role of forced labour in the Soviet economy also declined. It is unlikely that penal labour now plays a significant role in the forest industries. In the 1930s it would seem that if camp labour was economic, it was only marginally so. Clearly, camps had to be built, prisoners had to be fed (even if the rations were low) and guards had to be paid, fed and accommodated. But today the economics of forced labour are likely to be even less attractive. In order to compete with free labour the output from any camp would have to be as high as possible; this would necessitate the use of a far higher degree of mechanisation than was needed in the 1930s, in fact it would probably require a level of mechanisation not far removed from that in common use. So unless a fairly high level of mechanisation is used, forced labour is unlikely to be economic, but on the other hand, it is unlikely that the authorities would consider the use of new, expensive machines by convict labour. So it would seem that unless the political climate alters radically, there would be no incentive to use penal labour again on the scale that it was used in the 1930s.

Institutional Factors and Labour Mobility

As early as 1920 Trotsky was demanding that an individual's freedom to choose his place of work should be abandoned. This was not supported by others, but in 1922 the Labour Code for the RSFSR stated that, with certain exceptions, citizens could be called upon to do work in the form of a compulsory labour service. One of the major problems that the Soviet Union faced in the 1920s and 1930s, which led them to adopt a labour policy that tended towards the direction of labour to various jobs, was the high labour turnover that was a feature of industry. In 1930, for example, figures suggest that, on average, coal-workers changed jobs three times a year (Nove, 1969). In 1932 an internal passport system had been introduced for urban dwellers over the age of sixteen; this was to control the movement of the population because the passport was necessary to move into and between urban communities. In 1938 the Labour Book system was introduced to attempt to reduce labour turnover; every worker was obliged by law to carry a 'work book' which showed his record of employment as well as the reasons for his release from various jobs. But even before the introduction of this formal record of a worker's life, moving often from one job to another, or absenteeism was combatted by discriminatory action such as withdrawal of vacation rights, etc.

In 1940 as war began to appear imminent, workers were actually forbidden to leave any enterprise or institution arbitrarily. Ministries were empowered to move certain of their work-force such as skilled workers and technicians, from enterprise to enterprise. Absenteeism or lateness of more than twenty minutes was to be severely punished. This particular law remained on the books until 1956, although it was not enforced after 1952.

Apart from these legal methods of directing labour, institutions exist which lie somewhere between the persuasive and legal methods of controlling the labour supply. In the 1930s *Orgnabor* was set up and this concerned itself mainly with transferring rural workers into the growing urban and industrial centres. After the war, *Orgnabor* occupied itself largely with transferring labour from the western parts of the Soviet Union to the developing eastern areas. Also, the *Komsomol* (The Communist Youth League) exerts a 'persuasive' influence on the younger work-force. The Baikal-Amur Railway was declared to be an 'all-union *Komsomol* crash project.' The *Komsomol* in this situation presumably exerts some moral persuasion on its members to volunteer for work on such projects.

One area of direction which still exists that is of importance to forestry is the legal power that the Government possesses to direct graduates from establishments of higher education. Graduates from technical-vocational schools and full-time students from specialised higher educational institutions, and this includes institutions specialising in forestry, are assigned to specific jobs for a period of two to four years. This aids the recruitment of qualified staff in the less desirable areas of the Soviet Union such as the north and east. It is likely that without this direction recruitment of such labour in these areas would be very difficult indeed. However, often this system does not work satisfactorily; sometimes, for example, lists of jobs are circulated amongst the graduating students with the best students receiving the lists and making their choices first. This means that the least desirable jobs (arguably those in the areas of the most importance to the future production of *Minlesbumprom*) are filled with the least able students who make their choices last of all. Although the wishes of the students involved are supposed to be consulted, most authors state that many students are not satisfied with the jobs they receive and that the enforcement of this particular set of regulations is often very difficult.

PRESENT PROBLEMS OF THE LABOUR SUPPLY

With the overwhelming evidence now in the West of the use of penal labour in the Soviet Union in the 1930s, as well as the accounts of collectivisation, it is sometimes thought that labour policy in the Soviet Union was little more than forcing workers to move to the work. There is a temptation to assume that this is still the case today, and so assume that problems of labour supply to various industries can be ignored. However, Article 40 of the new constitution states that citizens of the Soviet Union have the right to work along with the right to choose their profession in accordance, of course, with educational qualifications. Helgeson (1979), in a recent small paper, concluded that 'we can no longer speak of the Soviet Union moving labour ... but only of attracting labour'.

If what Helgeson says is true the forest industries must compete for their work-force with other industries. From this it follows that the supply of labour to the forest industries will depend upon, amongst other things, the relative attractiveness of work in forestry. It is here, as will be seen, that the forest industries may face a severe test.

It is well known that the Soviet Union has, in recent years, been suffering from a slow-down in economic growth; part of the reason for this is the decline in the rate of growth of the labour force. The emphasis in the tenth five-year plan on 'quality and effectiveness' derived, in part, from the realisation that the ability of the Soviet Union's economy to grow extensively was coming to an end. The possibilities of creating economic growth by investment in new enterprises and attracting or diverting labour from the agricultural sector to new industry have come to an end, and with it the relative ease with which high growth rates can be achieved. Now intensive growth is needed. The resources available to the Soviet Union, especially labour, have to be employed in their best possible uses and have the maximum output 'squeezed' from them.

Some might argue that it is in this type of situation that a Soviet style planned economy would be least effective. Strong central control of all of the factors of production may be able to produce high growth rates in the economy during the early stages of development but later, it may be argued, with the development of a more complex economy producing many different goods and presenting many possible investment opportunities, the market system with its variable prices and greater reliance on private initiative is better equipped to 'micro-economise' or ensure that the maximum value is extracted from all of the factors of production available. The lack of flexibility associated with strong central control may mean that investment opportunities remain unrealised until they are brought to the notice of the planners. In this type of environment it may be that labour becomes the most mobile of the factors of production. With capital, equipment and land under direct control it may be that labour adjusts more quickly to relative shortages and gluts than other factors of production. As the demand for labour, relative to the supply, increases it follows that there will be more opportunity for workers to choose the type of employment that they want rather than accepting jobs which are their second or third choices. Therefore, it would be expected in the future that forestry will have to rely less on the fact that there is a relatively large number of workers looking for jobs and more on its own merits as a job in comparison with jobs in other parts of industry.

In this situation if the forest industries are not successful in this respect part of the productivity growth which the new machines described in the last chapter allow will be 'absorbed' as the number of workers employed in forestry declines. Indeed, it was this potential shortage of labour that caused much of the impetus to go into the development of new, highly productive logging machines. But before discussing forestry's relative attractiveness to workers when competing for labour it is useful to begin by looking at the overall size of the pool of labour within which forestry must compete.

Population Growth

The Soviet population in mid-1978 has been estimated as being 261.4 million (Grossman, 1978). Since the mid-1960s the growth rate of the Soviet population has fallen from its previous level of about

1.5 per cent per annum to less than 1 per cent. In Table 6.2 average annual percentage growth rates of population are shown. The figures for 1980 to 1990 are, of course, predictions. As may be seen, the growth rate of population in the Soviet Union as a whole is expected to stay approximately constant at around 1.0 to 0.8 per cent, but for forestry the breakdown of the figures into individual areas highlights a major problem. The growth of population is not expected to be evenly distributed about the Soviet Union. In the poorer, central Asian Republics and in the Caucasus area the estimated rates of population growth are quite high. In the RSFSR, however, the expected rate of growth is very low. This is, of course, the reverse of the pattern of the distribution of forests and forest industries in the country. About 90 per cent of the timber extracted in the Soviet Union comes from the RSFSR - see Chapters 2 and 3. Most of the potential timber stocks are, in fact, situated in the northern European parts, Siberia and the Far East. The population growth figures are not broken down to show the extent to which population growth varies within the RSFSR, but there seems to be no good reason to assume that the growth rate in the heavily forested areas should be greater than or less than the overall Republic average.

TABLE 6.2 Population Growth Rates 1971 - 1990

	1971-75	1976-80	1981-85	1986-90
Soviet Union total	0.9	0.9	1.0	0.8
RSFSR	0.6	0.6	0.5	0.3
Ukraine	0.7	0.5	0.4	0.3
White Russia	0.9	0.6	0.8	0.6
Baltic Republics	0.9	0.5	0.3	0.3
Central Asian Republics	2.8	3.0	3.2	3.3
Kazakhstan	1.4	1.7	1.9	1.9
Caucasus	1.8	1.6	2.1	1.8

All figures are average annual percentage changes.

Source: Grossman, 1978

Working Age Population

The important figures for the economy are the variations in the population of working age. These will be related to the overall population changes but will also depend upon death rates within the working age population and the birth rates in the previous years. Table 6.3 shows the actual and expected increments to working age population in millions for the relevant five-year periods. The figures refer to males between the ages of 16 and 59 and females between the ages of 16 and 54. As may be seen from this table, the situation for forestry is worse than that suggested by Table 6.2

TABLE 6.3 Increments to the Working Age Population 1971 to 1990

	1971-75	1976-80	1981-85	1986-90
Soviet Union total	12.7	11.1	3.2	2.5
RSFSR	5.9	4.9	-0.3	-0.7
Ukraine	1.7	1.2	-0.1	neg
White Russia	0.5	0.4	0.1	neg
Baltic Republics	0.3	0.2	neg	neg
Central Asian Republics	2.0	2.3	2.0	2.0
Kazakhstan	1.1	1.1	0.7	0.6
Caucasus	1.0	1.1	0.7	0.5

Figures are in millions for the entire five-year periods. Figures may not sum to the total because of rounding errors. Neg = negligible

Source: Grossman, 1978

concerning population growth as a whole. Whilst for the whole of the Soviet Union the working age population is expected to increase by 5.7 millions between 1980 and 1990, for the RSFSR it is actually expected to decline by 1 million. This represents about 1 per cent of the total work-force in that area.

Possible responses. This reduction in the size of the work-force in a major industrial region, and the major logging area of the country poses long-term planning problems. There are several possible responses that the Soviet planners could make in this situation. The technological response has already been discussed as far as logging is concerned - an increase in labour productivity is called for. The reason why this particular option is emphasised in the Soviet technical press is that really there are no other responses to the labour shortage that are likely to be effective.

Already, the participation rates of the working age population are very high. In 1970, in the 20 to 59 year age range, 93.2 per cent of all males were economically active, and in the 20 to 54 year age range 88.9 per cent of all females were economically active. This compares with the 96.3 and 54.8 per cent respectively in the United Kingdom, and 90.9 and 50.4 for the United States (Grossman, 1978). So there seems little possibility of increasing participation rates by attracting more of those members of the population of the relevant ages into work.

Another possible response to the shortage of labour would be to re-allocate or to attract labour from the agricultural sector into the industrial sector. This has been a feature of the development of the Soviet economy, as it is in many developing economies, but now the potential for it is much less than it has been in the past. The percentage of the total work-force engaged in agriculture in the RSFSR and the western European part of the Soviet Union has fallen from 28 per cent in 1965 to 20 per cent in 1975. So there seems to be little real scope for a continued influx of workers from the agricultural sector into industry in the future. This conclusion is reinforced if one considers the performance of the agricultural sector in recent years. It is unlikely that planners concerned with agricultural output would view with favour the outflow of some of their work-force and would probably oppose any measures designed to encourage it.

Another factor that might be able to assist certain industries over the problems created by labour shortages is inter-regional migration. Here, however, the outlook for forestry is also depressing. Clearly migration will occur, institutional factors permitting, if the prospects of work in the area to be moved to are more attractive - in terms of both monetary and non-monetary factors - than work in the area from which the workers are moving. Unfortunately the north of European Russia, Siberia and the Far East do not fit these requirements. The climate is severe, the cultural facilities are significantly less developed than those elsewhere and the supply system is such that even apparent necessities such as meat are often in short supply in the shops in these areas. Also, as much of the possible source of labour lies in the 'surplus' areas of Soviet Central Asia and the Caucasus, where the climate and supply of food are often considerably better than those in the RSFSR, it is unlikely that the change from life in Georgia, say, to life in a logging town in Siberia will appeal to many.

Table 6.4 shows some figures for migration into the RSFSR, the Far East and Siberian parts of Russia, and southern Soviet Union. As can be seen, during 1959 to 1970, the period that the table refers to, the RSFSR had a net out-migration of about 2.1 million people. The Far East and Siberian areas of Russia, where most of the timber stock is situated and where logging operations are to be concentrated in the future, shows a slightly different pattern. Here there has been a net in-migration of 270,000 workers between the ages of 20 and 29, although the overall figure is one of out-migration. This may be interpreted as being a temporary movement of young mobile workers, moving to Siberia and the Far East to take advantage of the

TABLE 6.4 Net Migration 1959 to 1970

(Figures are in thousands and refer to the whole period)

Age Group	RSFSR	Siberia and the Far East	Southern USSR
10-19	-700	-355	+985
20-29	-29	+270	+124
30-39	-569	-277	+823
40-49	-272	-308	+475
50-79	-577	-382	+726
Total	-2,147	-1,003	+3,133

Source: Helgeson, 1979

higher salaries and wages offered there and to see a bit of the country, only to return later to other areas of the Soviet Union. As was discussed in Chapter 5, this is a situation that would not greatly benefit the logging industry in the future for the equipment being introduced requires experience and commitment on the part of its operators in order for its full productivity to be realised. Young, inexperienced, mobile workers are more likely to be less skilled in the use of such equipment so reducing productivity and reliability and increasing servicing and repair requirements.

The Potential Labour Force for Forestry

Forestry has to compete for its labour force in the pool of the working age population in industry - apart from some temporary workers that are drawn from the agricultural sector. Table 6.5 shows what has happened to the numbers employed in industry from 1961 to 1975. The figures are annual percentage changes in the numbers employed in various branches of industry.

TABLE 6.5 Average Annual Percentage Changes in the Numbers Employed in Industry

	1961-65	1966-70	1971-75
Industry total	3.9	2.9	1.5
Coal	0.1	-1.4	-2.7
Chemical industry	9.6	4.6	2.3
Iron and Steel	3.4	1.2	0.1
Engineering	6.6	3.9	2.8
Construction	1.7	3.1	1.5
Light industry	2.2	3.1	0.3
Forest Products Industries	0.9	0.2	-0.4

Source: Grossman, 1978

There has been a decline in the growth rate of industrial workers as a whole and this has been reflected in the decline in the growth rates of the number of employees in the various types of industry listed in the table. However, the figures can be used to show a shift in employment between the branches of industry. The chemical industries and engineering show growth rates greater than the overall level for industry as a whole implying a slow shift in employment into these areas. The coal and forest products industries as a whole show changes in employment consistently below the average indicating a movement away from these industries.

The figures in Table 6.5 refer to the forest products industries. Looking specifically at the logging industry, and using figures for the output of timber and the overall productivity of labour, a similar pattern emerges. The increases in the productivity of labour have been greater than the increases in actual output, implying a decrease in the size of the work-force. Between 1965 and 1975 the number of workers employed in logging and work at the lower landing declined by an average of 2 per cent per annum (Anon, 1977b). There are good reasons for interpreting these figures as illustrative of the fact that workers consider employment in logging and the forest products industries as less desirable than employment in other areas of the economy rather than as a voluntary shedding of labour by the forest industries.

If a 1 per cent growth in the industrial working population is the maximum that can be expected in the 1980s (Eason, 1979), the decline in the number of people working in forestry can be expected to continue and even accelerate. If the crude relationship between the rate of change in industrial workers and the rate of change in the number of employees in the forest product industries remains the same as that suggested in Table 6.5, a 1 per cent increase in the size of the industrial working population should be accompanied by a 0.7 per cent annual decline in the number of people engaged in wood conversion. By the simple extrapolation of the trend exhibited the work-force in logging can be expected to fall by 2 per cent per annum, or even slightly more in the future. If the nature of a career in the logging industry is considered (especially for someone not at the managerial level) this figure is not all that surprising.

Work in the Forest Industries

If forestry is going to have to compete with other industries for its labour force in the future it is necessary to get some idea of the appeal of a career in forestry compared to other branches of industry. There are two pieces of evidence concerning people's personal evaluation of life in forestry. First, there is some direct evidence of individual evaluation of the profession when directly compared to other jobs and secondly, there is the indirect evidence of labour turnover which is of interest both for its own sake, and also because of the implications that it has for the future recruiting needs of the forest industries.

Direct evidence. V.N. Shumbkin (quoted in Loiberg, 1978) undertook research to try to ascertain directly people's opinions of a career in forestry. He surveyed school-leavers in four areas and his subjects came from large towns, from medium-sized towns, from small towns and finally from small villages including logging villages. Students were asked to rank a number of professions according to their desirability, the maximum possible mark that any profession could score being 10 and this represented maximum desirability. When the results were analysed it was seen that a career in the forest industries was ranked bottom. The results of his survey are tabulated in Table 6.6. Forestry scored little more than 3 points, doing neither noticeably worse nor better in the answers from students from small villages who should, arguably, have been better informed of the realities of a job in forestry. Surprisingly, for the author at least, both mining and working in the iron and steel industry scored noticeably higher points than logging.

TABLE 6.6 Results of a Survey of School-Leavers

	School-leavers from			
	large towns	medium towns	small towns	villages
Logging worker	3.01	3.45	3.13	3.15
Engineering worker	4.86	5.44	5.46	5.24
Miner	5.18	6.17	5.79	5.43
Iron and steel worker	5.42	6.24	6.75	5.60

Source: Loiberg, 1978

Reasons for the low score in forestry are not given in the results of the survey. However, given the wage differential that exists, one might speculate on the degree of manual labour required in the forest industries, the lack of appeal of an 'outdoor life' in the Siberian conditions and the desire of most people to move, if possible, to the larger towns where living conditions are better as being the reasons for the poor showing of logging. There is evidence that these reasons are of importance in that they are often given as reasons for a worker leaving a job in forestry. One other reason, which is unlikely to be commented upon by the Soviet press, is one which emigré Russians have mentioned to the author. This is that possibly as a result of Stalin's labour camps, forestry is thought of as being a job for criminals or people who can get no other employment; it is not really considered as being a job for a 'respectable' person. This may be true in the case of the older generation who can still remember such things, but for young people there is probably little truth in this view. The same emigrés who believe that forestry has such a reputation also state that mining is in the same category, but this career scored well in the survey of school-leavers. It is more likely that the younger generation in the Soviet Union is swayed by more materialistic considerations such as living conditions and the ease with which a job can be performed.

Indirect evidence. An indicator that is often considered to show the relative desirability of a particular job is the rate of labour turnover. If labour turnover is high it may be expected that the job is unsatisfactory from the workers' point of view. Overall, the turnover of workers in the logging industry has fallen in recent years from 31.8 per cent in 1965 to 24.1 in 1975. These figures express the number of workers leaving forestry as a percentage of the average numbers of workers involved in forestry during the year. However, the figures are more disturbing if looked at in a disaggregated form. Labour turnover is 1½ times lower in such areas as Georgia and White Russia, where living conditions might be expected to be better. But in Siberia and the Far East, the important timber areas, it is considerably greater and in parts of *Dalesprom* and *Zabaikalles* the figure exceeds 100 per cent (Dmitrin, 1978).

Of the people leaving forestry in 1975, one-fifth complained of the poor living conditions and lack of facilities associated with a career in forestry and the concomitant necessity of living in remote areas, and nearly a third left for reasons concerning the nature of work in the logging industry, the heavy physical nature of the job (Traktinski, 1977).

The labour turnover varies significantly between different age groups; for workers between the ages of 18 and 24 the figure, according to information for 1972, was 40.7 per cent. For ages between 25 and 29 it was 44.4 per cent; for ages between 30 and 39 it was 27.9 per cent; 40 to 49, 16.7 per cent and 50 to 60, 14.9. This shows that in the important age range of up to 30 years of age no less than 40 per cent of the workers leave each year. Referring to the high figures for all ages in the Far East and Siberia, the turnover of labour in this age group in that area must be extremely high. This tends to reinforce the view that the net in-migration into Siberia and the Far East in the 20 to 29 age range shown in Table 6.4 is really a manifestation of an itinerant element in the work-force of that age, and probably not an expression of a part of the labour force wanting to settle permanently in that area.

But these figures have another significance. The average age of a worker leaving the forest industries is 32.6 years; however, the average age of a worker entering the industry is 34.4. This means that, progressively, the work-force in forestry is becoming older. In 1972 the average age of a worker in logging was 37 and in all probability this 'average worker' has become older by now (Loiberg, 1978). Even so, this figure was two years older than the average industrial worker in the Soviet Union. As the average age of the work-force increases this implies a skewed distribution in the age classes within the profession. Thus there will be a greater number of workers in the older age groups compared to the younger ones. This means that, in the future, a larger number of workers will begin to retire and, thus, the recruitment rate must be correspondingly increased in order to maintain the size of the work-force at a constant level. Already, in certain operations, the average age is greater than the overall average of 37 years. The average age of a timber feller is more than 40 years, for a delimber it is about 38 years and for a person cross-cutting timber at the lower landing about 40 (Loiberg, 1978). Looking at the case of a timber feller, the pension age in this branch of forestry is 55; ideally, if the youngest feller is 18 years of age, the average of fellers in logging should be 36½. If this were the case there could be a constant retirement rate and a constant recruitment rate and the average age would remain the same through time. However, the evidence suggests that fellers rarely work at felling past the age of 48 or 50 and so the average age of 40 implies an even greater skewed distribution of the ages of fellers than appears at first sight.

All of this evidence suggests that the forest industries are likely to face severe problems of maintaining the work-force at the required level in the future. As these figures for workers' ages are now old (the article which gives them was published in 1978 and the author of that article was referring to research done before that) these problems must now be becoming more acute. In order to keep the numbers employed in forestry constant, the rate of recruitment must be increased. But all that has gone before would tend to lead to the conclusion that, in competition with other branches of industry, forestry will not fare well in the face of a declining growth rate of the working age population.

Also, if the older work-force members are replaced by younger workers, the overall average productivity level of the industry will decline. Some authors claim that the greater mobility of the younger workers, quite apart from their lower levels of skill, reduces their own feelings of responsibility to the profession and so reduces their productivity. For whatever reason, according to one estimate one-quarter of the workers under the age of 30 fail to reach the work norms set compared to only about 18 per cent for workers over 40 (Loiberg, 1978).

Conclusions

The figures for the changes in the number of people employed in the forest industries in Table 6.5 might be open to the interpretation that many people were clammering for a job in forestry in the open air and that the changes observed were the result of forestry being able to employ only the numbers they required to fulfil the plan. However, the tone of the articles in the Soviet press and the evidence presented concerning the assessment that people made of a profession in forestry invalidate this interpretation. With machines standing idle for want of servicing and the plans for road building and, sometimes even production, going unfulfilled, it is likely that the forest industries would like to employ more rather than less personnel. If this is the case, the figures must be interpreted, as they have been in the text, as being the best that the industry can do.

In this situation, the decline in the growth rate of the working age population is indeed worrying. Earlier it was stated that by extrapolating past trends into the future, a decline in the work-force in logging of 2 per cent per annum or more might be expected. But as well as this there is the argument just put forward, that the retirement rate is likely to rise in the future. With this effect, it is probable that the recruitment rate will fail to match the retirement rate so that there will be a decline in the quality as well as the quantity of the work-force.

Recalling the conclusion to the previous chapter, it was shown that the productivity of labour would rise in the future by between 1.5 and 3.1 per cent per annum, with the likelihood that the lower figure would turn out to be true. If the size of the work-force in logging is to decline by 2 per cent or more each year, we are now in a position to predict the likely output from the Soviet Union's forests. Using the simple relationship that the output of timber is equal to the productivity of labour multiplied by the size of the work-force the increase in the volume of timber logged is only going to be at the most 1.1 per cent per annum - assuming an increase in the productivity of labour of 3.1 per cent. There is the possibility that output will in fact decline by up to .5 per cent per annum.

If the optimistic prognosis is correct the output of timber from the Soviet Union's forests in the year 2000 will be about 430 million cubic metres, compared to the planned figure for 1981 of 342.7 million (Kirin, 1981). As the increase in labour productivity is likely to be nearer the lower level than the upper limit, the output of 430 million cubic metres is the maximum that is going to be achieved. In all probability the output of timber will not grow appreciably in the nearer future especially and 400 million by the year 2000 is probably a closer estimate of what will occur.

Some of the responses to this situation will be discussed in Chapters 9 and 10. In the next two chapters the planning system and some economic problems will be examined and an attempt will be made to associate these aspects of the Soviet system to the performance of the forest sector. But before that, other aspects of the labour supply in forestry will be looked at.

WAGES AND WORKING CONDITIONS

Whilst the mechanisation of operations in the forest industries can do its part in the improvement of the actual job of working in forestry and so help to increase the number of people who would like to work in the industry, the working conditions and wages play an important part in attracting people to the profession.

Working Conditions

There have been two common methods of allocating the 41 hours of the working week in use in the logging industry until recently. First, during the spring, summer and early autumn when the hours of daylight are fairly long, eight-hour days are worked for five days and every eighth week an eight-hour shift is worked on Saturday for normal pay. In winter, when the hours of daylight are less a seven-hour day is worked for five days of the week and a six-hour day every Saturday.

With the introduction of the new technology, the higher rates associated with amortisation payments and other costs of using the machines have induced many *lespromkhozi* to introduce shift working. In shift work the day shift works an eight-hour shift six days a week, the night shift a seven-hour shift five days a week and the evening shift an eight-hour shift for five days. This probably represents a reduction in the quality of work in forestry from the point of view of a potential worker.

Most workers are entitled to twenty-four working days holiday, although temporary and seasonal workers such as those who work at timber floating in the season are entitled to none. In an attempt to reduce labour turnover, workers who have worked at a concern for an uninterrupted three years are entitled to an extra twenty-four days holiday, giving forty-eight days each year. In general workers in the far north receive extra holidays but such arrangements do not usually affect forestry concerns.

Apart from the usual breaks during the day, workers are allowed extra breaks during cold weather - sometimes up to 20 minutes per hour; details are usually negotiated locally. The workers who work in the open in the forests or lower landings are issued with special clothing - jackets, trousers, boots and hats; times of issue of these are agreed locally and they are only to be used at work

(Bronina and Chubais, 1969).

Many women work in forestry in the Soviet Union, many delimbing trees with axes, others work at lower landings. There are several labour laws that affect them: they are not allowed to work at the heaviest of the work at logging concerns, and night work is impermissible except as a temporary expedient. Pregnant women and those with children of less than one year of age, if they are still working, can, on doctor's orders, move to lighter work with no loss of pay. Fifty-six days before and after the birth are allowed off with payments during that time from social insurance, and women's jobs are secure for one year if they wish to take time off for that period after the birth of a child. This year is without pay, however, (Andreyev, 1976). The high participation rate of women in the Soviet Union quoted earlier shows that women tend to work more after having children and spend less time at home bringing them up. There is, thus, a considerable investment in nurseries for children either as general nurseries in the town or as specific ones for forest enterprises. Also, it is very common for a married couple to live with one of the in-laws until they receive their own flat - which can be a matter of years after marriage. It is usual for the grand-parents to look after the children whilst both of the parents work.

The Wage System

For blue-collar workers in forestry there are two methods of calculating the wage. Payment may be either on an hourly basis or the worker may operate on a piece work basis.

For hourly payment, the hourly wage will be equal to Wh in the equation below. In this equation TR is the tariff rate, TF the

$$Wh = TR \times TF \times RF$$

tariff factor and RF is the regional factor. The tariff rate, TR, is the basic wage rate for the operation in question. The rates for 1977 are shown in Table 6.7. The tariff factor, TF, represents an adjustment to the basic wage rate for skill and experience. The figures for the tariff rates shown in Table 6.7 relate to the lowest skilled workers or those with the least experience. For such workers the tariff factor is unity. There are other tariff factors which relate to workers of greater skill; these are 1.10, 1.22, 1.38, 1.58 and 1.85. The regional factor, RF, represents an attempt to attract labour into the less attractive areas of the country. The regional factors in use are 1.0, 1.15, 1.2, 1.3, 1.4, 1.5, 1.6, 1.7, 1.8 and 2.0. Regional factors increase as the conditions of work deteriorate and a factor of 2 is used in areas north of the Arctic Circle. Factors that are most common in forestry concerns are 1.15, 1.2 and 1.3.

The piece work system operates in a similar manner to the hourly system. The wage per unit of output, Wq, is defined in the equation below where N is the hourly work norm. Usually the tariff rate for

$$Wq = \frac{TR \times TF \times RF}{N}$$

piece workers is somewhat higher than that used for hourly paid workers. The tariff rate for workers on the forest floor working

TABLE 6.7 Tariff Rates for Various Operations

All operations on the forest floor	47.8 kopecks per hour
Other operations at the lower landing	40.4 kopecks per hour
Other operations at the lower landing under bad working conditions	45.6 kopecks per hour
Repair personnel	41.8 kopecks per hour
Road building and road maintenance	43.8 kopecks per hour

Source: Rodigin, 1977b

under a piece work system is 51.2 kopecks per hour compared to the 47.8 for hourly paid workers given in Table 6.7. Thus a feller with a work norm of 50 cubic metres per seven-hour day, working in an area with a regional factor of 1.3 and with a personal tariff factor of 1.38 would receive 12.9 kopecks per cubic metre felled.

The piece work system is thought of as being the most useful in the forest industries because of the fact that it creates a greater interest and incentive for the worker to produce as much as possible. When the emphasis is on quality rather than quantity such a system of wage payment may not produce the desired results and so an hourly wage system would be employed. In forestry, however, such a situation is rarely encountered and the piece work system has been modified to encourage an even greater interest in the quantity produced. Whilst only 10 per cent of the workers in forestry are paid according to the straightforward piece work system as described by the equation above, 70 per cent are paid by a slightly modified piece work system where any extra output over and above the norm is rewarded at a higher rate of pay. For workers on the forest floor and for those extracting timber from the upper to lower landing there is a 20 per cent bonus for actually fulfilling the norm - usually calculated on a monthly or half-monthly basis - and an extra 2 per cent for every 1 per cent above the norm that they manage. For workers at the lower landing the figures are 15 and 1.5 per cent respectively. Brigade leaders are paid an extra 10 to 15 per cent depending upon the size of the brigade. The bonuses are usually paid only if the whole brigade fulfils the norm; this puts pressure on all of the members of the brigade and ensures a form of self-policing. These bonuses are paid out of the wages fund and are distinct from bonuses paid from the various funds for economic incentives that are built into the *tekhpromfinplan*. These will be discussed in Chapter 7.

In competing with other industries there would appear to be little room for manipulation. The tariff rates are fixed by national agreement as are the various factors applied to find the overall wage level. However, there is some latitude in the system. In the early 1970s bonuses were increased for people with long service and recently a new tariff factor of 2.11 was introduced for drivers of feller-bunchers and feller-skidders. Also there is some leeway in defining the norm levels. Although basic norms are defined at a national level, some modification is permissible at local level, so setting relatively easy norms would allow the effective wage rate to be increased. Also, the tariff factor which is defined according to the level of experience is an area where something can be done. It is quite common in the Soviet Union to respond to a shortage of labour by 'over-grading' workers. Already it has become common in forestry to regard tariff factor 1.10 as referring to the lowest grade - so increasing the overall wage rate (Petrov, 1971a).

However, it is probably true to say that, in general, in the Soviet Union it is not the money wage rate that is of primary concern but the real rate in terms of the goods that can be purchased. There is a shortage of many items for consumer expenditure and so the question of the availability of goods is probably more important than the overall wage. Attracting workers into forestry in the future probably has less to do with the wage level than on the living standard in the important Siberian part of the country. Already people complain that those working in Siberia earn too much and are capable of saving sufficient that they can buy all the goods they need when coming into the large towns on holidays. The influx of young people into the Siberian area and the net outflow of other age groups is probably best explained by high wages but bad living conditions attracting the young for only enough time to make some money and move out.

Women in Forestry

The high participation rates of women in the labour market has been referred to earlier; this is reflected in forestry with a large number of women working in the forest industries. Table 6.8 shows the position in 1976 in *Archangelsklesprom*, which employed more than 30,000 thousand women - over 35 per cent of the total workforce.

TABLE 6.8 Women Workers in Archangelsklesprom in 1976

	Total number of workers	Number of women	Percentage
Working on the forest floor	20,178	3,450	17.1
including the number engaged in hard manual labour	7,927	3,352	42.3
Working at the lower landing	14,057	3,815	27.1
including the number engaged in hard manual labour	1,447	820	56.7
Total workers engaged in basic production	51,675	11,626	22.5

Source: Zheltov, 1977

Looking at the figures in Table 6.8 it is evident that women are concentrated in the harder, and generally less skilled, operations of production. Thus, almost all of the 3,450 women employed on the forest floor by *Archangelsklesprom* were engaged at hard manual labour such as delimbing of trees by hand, clearing snow from around trees in winter, helping in the loading of lorries, etc. Also, at the lower landing a far greater proportion of those engaged in the harder jobs, stacking, loading, sorting etc., were women. Although these women would be likely to have lower tariff factors because of their probable lower qualifications in comparison to the more skilled professions, the tariff rate for women is often adjusted upwards by between 6 and 10 per cent.

One of the effects of the introduction of new technology is that many of the operations which women tend to be concentrated in are being mechanised or are disappearing. For example, the mechanisation of the delimbing process will effectively reduce the requirement for women workers. New felling machines will reduce the need for snow clearing in winter around the base of trees and the mechanisation of sorting, loading and unloading at the lower landing will have a similar effect (Voronitsin, 1975).

Usually it is recommended that women are moved when mechanisation is introduced into lighter jobs, either still within the sphere of direct production or, more often, to other jobs within the *lespromkhoz* in the area of the various services that are provided to the workers. However, there are increasing numbers of women being trained for the more skilled jobs - mostly line operators at lower landings and repair personnel. More rarely they train as operators of feller-bunchers, feller-skidders and delimbing machines. In the late 1970s there were more than 7,500 female crane drivers and operators of saw lines at lower landings and in some of the training schools up to half of the workers being trained are women (Filippskikh, 1978 and Novokharstkin, 1978).

Industrial Injury

One of the aspects of the forest industries that reduces the appeal of the job apart from the fact that it is relatively hard work is the fact that it is relatively dangerous. In most countries the incidence of industrial injury in logging and the wood processing industries is greater proportionately than in other sectors. If the concerns that make up *Chelyables* are typical, up to 2 per cent of the logging costs of *Minlesbumprom* can be attributed to losses and expenditures associated with industrial injury (Guleka, 1970; Elenski and Korotishev, 1976). The figures published about the incidence of injury in forestry, although very incomplete, do tend to uphold the view that the increased mechanisation of the industry is leading to a reduction in injury. Between 1975 and 1980 the number of industrial injuries fell by nearly 30 per cent (Maklukov, 1982). This is one of the beneficial side effects of mechanisation and it is hoped that it will lead to forestry being looked upon as a more desirable career by its potential work-force.

Education in Forestry

Higher education in the Soviet Union in forestry is carried on in several institutions. Some, like the Kirov Forestry Academy in Leningrad, the Forest Technical Institutes in Moscow, Archangelsk and the Urals come under the control of the Ministry of Higher Education. The Kirov Forestry Academy, where the author studied, usually has 4,500 students enrolled for five- or three-year courses in many subjects. It is the largest institution of higher education in forestry in the Soviet Union (and probably the world) and is divided into departments dealing with logging mechanisation, economics, timber floating, woodworking and many other areas of the forest industries.

As well as these, the research institutions such as *TsNIIME*, *TsNIILesosplav*, *Giprolestrans* and *LenNIILKh* also accept students, usually for research in a particular field. These institutions come under the control of the Timber Industries' Ministry.

The first group of institutions in the control of the Ministry of Higher Education enroll about 2,000 students per annum and the second group enroll about 1,500 (Dmitrin, 1978). At the present moment about 70,000 specialists with qualifications from these institutes of higher education are working in logging, woodworking and the wood processing industries; this represents only about 30 per cent of all those personnel who are in jobs where such degree level qualifications would be relevant or desirable.

However, it is possible that the most important area of education is at the lower levels of the operations in the forest industries. At the present moment there is some dissatisfaction with the level of education of forestry workers. In the near future it is hoped to change this position and to increase the overall level of education and training of the workers - either of a general kind or specific to particular operations. The productivity of a person who has been on some sort of training course is often 20 to 25 per cent more compared to those whose education ended at secondary school (Traktinski, 1977). Thus it is hoped to increase the overal productivity of labour and also reduce the level of labour turnover by making the job more skilled and less of the type of job associated in peoples' minds with unskilled labour.

To this end it is intended to increase both the overall education received by such workers during training, including such courses as economics, politics, etc., and at the same time reduce the number of specialist courses taught at the colleges engaged in such training. At the present moment these schools tend to train workers for one or more of up to twenty professions in the forest industries and this, it is considered, spreads their individual resources too thinly (Khomenka, 1978 and Traktinski, 1977). Thus, at all colleges and schools in the various *obyedineniye*, there will be a general series of courses to increase the students' overall level of political and economic awareness but at the same time each school will prepare its students for one or more of only about eight professions.

In the mid-1970s a single central plan for this was set up in an attempt to co-ordinate this type of training and prevent wasteful duplication of courses. One part of the plan was to increase the number of professional-technical colleges. At present there are about ninety such institutions in operation with a capacity of more than 31,000 students. However, the construction programme is not going ahead as planned. In 1976 the overall plan for building such colleges was only 46 per cent carried out and in some *obyedineniye* the shortfall was even worse. In 1980 the shortfall in capital spending in this area was 38 per cent (Prodaivoda, 1981). In the words of one Soviet author 'the construction of professional-technical colleges should outstrip the construction of new logging concerns' (Traktinski, 1977). This is probably a very valid point as the failure to provide a trained work-force to operate the new technology could mean that the present investment programme becomes an expensive failure. For example, workers for feller-bunchers should ideally undergo 20 to 25 hours of supervised practice to train for operations in the forest. In fact they often get no more than 2 to 3 hours. So 'it is no wonder that they use twice as many spare parts in their first year' (Zverev, 1978).

Logging Towns

The shift of the emphasis of logging into ever more remote areas has meant the establishment of logging camps and towns has become necessary. Some of the camps have been mentioned in Chapter 3 in the discussion of the *vakhtovi* method of logging. In this system, where temporary camps are set up to house the workers for short periods of time, the main problems are the supply of prefabricated dormitories and other buildings. However, food and drink is also required and often hot meals are ferried in by helicopter for the canteen at the site.

In the more general sense, however, most Soviet experts would agree that the establishment of good living standards in the main logging villages and towns for the workers and their families is necessary if there is to be a reduction in the rate of labour turnover and an increase in the recruitment rate. However, this means that not only must there be investment in accommodation in small houses or small blocks of flats but also shops, canteens-cum-restaurants, proper roads and pavements and even hospitals will have to be built. In the larger towns as well, schools and nurseries will have to be set up for the workers' children. If this is not done, and if accommodation of the cruder hostel type is provided, the possibility of attracting married men and women who would be a more stable work-force is much reduced. Only the young and mobile will be recruited.

In the area around the Ust-Ilimsk logging combines four small logging villages are being built. These are at *Novaya Igirma* with a population of 2,000, at *Rudogorsk* with a population of 4,500, at *Tuba* with a population of 3,200 and at *Dalni* - population 1,800. Altogether it is expected that these will need a total of 108,980 square metres of living accommodation; this is about 9.5 square metres per person (Krupnik, 1975). In *Novaya Igirma*, for example, four-storey flats are to be built around the small shopping centre and are to be equipped with central heating, electric ovens and a centralised system of supply for radio and television programmes.

In *Vologdalesprom* during the period of the ninth five-year plan, 1971 to 1975, 1,700 families of workers in logging and woodworking received new flats, 10,000 moved into flats that had been refurbished, six schools were built with 1,960 places, five workers' clubs and six nurseries were built (Nikulichev, 1975). For the Ministry overall in 1974 over 400,000 square metres of new flats and houses for loggers were built and 1,300 thousand square metres were refurbished. All of this cost 165 million roubles (Terentyev and Soldatov, 1975).

However, the situation in the area of investment in accommodation for workers is probably unsatisfactory. This is a difficult area to be precise about. The construction side of the logging industry has almost consistently under-fulfilled its plans. In 1979 the plan for all building was only fulfilled by 78 per cent, part of this plan was the building of over half a million square metres of living accommodation and schools for 3,700 children (Batrakov, 1980). This is partly due to the way in which building is planned (in terms of building starts rather than completed buildings); see the next chapter for a more detailed discussion of this. The repair of the living accommodation is also an area of importance (if the experience of the author in living in the country is anything to go by) but it is also an area where plans are often under-fulfilled. Recently, for example, the repair plans for *Dalesprom*, *Omskles* and *Tomlesprom* were only 30 to 40 per cent carried out (*Lesnaya Promishlennost*, 11.12.80). Also it is probably true that the supply of foodstuffs is worse in the small villages in remote areas than in the larger towns. This, even if the accommodation is satisfactory, will reduce the attractiveness of life in the remoter areas of the country. It is likely that the forest industries will have to do a lot if they are to reduce labour turnover by presenting a more attractive face to the potential work-force.

7
The Economic System and Forestry

INTRODUCTION

So far the discussion of the logging industry has skirted around the fact that the concerns involved in forestry in the Soviet Union are embedded in a very different economic system to that which Western foresters find themselves acting within. The Soviet economy is a planned one. The term 'planned' in the Soviet Union is used not only as a description of some conscious process of determining outputs and inputs, but also as a word suggesting more rationality and, by implication, more desirability than the market system which is sometimes referred to as 'the chaos of the market'. In this and the next chapter the economic system will be examined from the point of view of forestry and the forest industries. In this chapter two main areas will be dealt with; these are the planning process itself and the system of price formation. In the next chapter some individual economic problems will be looked at. The wage system, itself a very important part of the economic system, has already been dealt with in Chapter 6.

THE PLANNING SYSTEM

Introduction

The most well-known plan that is usually associated with the Soviet Union is the five-year plan. Longer period plans exist - one was examined in Chapter 5 - but these plans are usually more tentative in nature and do not carry as much force as does the five-year plan. The tenth five-year plan, from 1976 to 1980, emphasised the more intensive use of resources and the increase in the productivity of labour by 25 per cent over the five years in question. The outcome of the plan was not very satisfactory; the productivity of labour did not grow by the planned amount, some even report that it fell (Solomonov, 1981). The output of timber from the Ministry's forests was less than the amount felled and extracted in 1970 (*Lesnaya Promishlennost*, 7.2.81). The eleventh five-year plan looks for an increase in labour productivity of between 16 and 18 per cent over the five-year period and in 1981, the first year of the plan, the output of timber from *Minlesbumprom's* forests was expected to be 222.5 million cubic metres. This is about 20 million more than was harvested in the last years of the tenth five-year plan. In Chapter 5 it was seen that the plan for increases in labour productivity is probably optimistic.

However, as a system of planning at the operational level, the five-year plan lacks immediacy. Being given a target to fulfil in five years' time is hardly an incentive to work hard in January of the first year of the plan's operation. The five-year plan is, therefore, broken down into one-year plans that are used at the enterprise level to plan production and to assess performance. The one-year plan is called either the *tekhpromfinplan*, this for *lespromkhozi* or enterprises with some physical, saleable output, or the *proizvodstvenna-finansovi* plan for *leskhozi* or enterprises engaged in sylvicultural operations from which no large amount of physical output directly derives.

The *tekhpromfinplan*, which is derived from the words for technical, production and financial plan, is the plan which a *lespromkhoz* will try to conform to or better during the year in question. It is no lightweight document, consisting of between 160 and 200 pages. These pages contain figures referring to the year's production and define the planned amounts of production as well as going into such detail as the amount of fuel to be used by the tractors and chain saws in the enterprise.

Clearly, the centralised planning of such detailed plans for each *lesopunkt* in the whole of the Soviet Union, as well as for each productive unit in all other industries, on an annual basis, is beyond the capabilities of *Gosplan*, the State Planning Agency. So a process of planning that involves the formation of suggested plans by each *lespromkhoz* is used.

The Planning Process

The production of a *tekhpromfinplan* for a given year for a *lespromkhoz* begins with the five-year plan which contains certain general indicators that have to be fulfilled. These figures refer to the level of production, the employment and payment of labour, the costs of production and profitability and the main elements of the finances of an enterprise. These figures are used as a basis for the production of the one-year plans which, clearly, have to be consistent with the overall aims of the five-year plan.

Formerly, the director of a *lespromkhoz* or *leskhoz* received certain 'control figures' in about September of the year before the year that the plan was to refer to. These were passed down to him from the Ministry. These 'control figures' detailed the main indicators that an enterprise should aim for and were broadly similar to those outlined above as contained in the five-year plan. The 'control figures' were worked out between the various ministries and *Gosplan* with the latter acting as a 'balancer' of the various plans so that the inputs required for the industries to produce the amount needed of them were consistent with the outputs of the industries producing raw materials etc. Now, however, this stage appears to have been dropped, in forestry at least, and so the planning process begins with the director drawing up a suggested plan for his enterprise of the basis of the figures contained in the five-year plan. This plan is less detailed than the final 160 to 200 page document and is worked out mainly on the basis of fixed coefficients of production (Rodigin, 1977a,b,c and d).

For example, Table 7.1 shows how the requirements for skidding tractors are calculated on the basis of an output level of 490 thousand cubic metres from a *lespromkhoz* in one year and the requirement for payment to the people working at repairing the skidders. In the table the figures asterisked once (*) were taken from tables of norms and so are out of the direct control of the individual director of the *lespromkhoz*. Figures marked with a double asterisk (**) are capable of being manipulated within the enterprise by agreement between the trades unions and the management. The other figures derive directly from the asterisked numbers. All of the other elements of the whole plan are worked out in a similar manner.

TABLE 7.1 Planning the Use and Repair Costs of Skidders

Volume to be harvested	490 cubic metres
Existing norm for a TDT-55 per machine shift*	57.0 cubic metres
Norm locally agreed**	59.8 cubic metres
Requirements of machine-shifts	8,200
Shifts required for other work, e.g. roading*	2,200
Total requirements of machine-shifts	10,400

Shifts per day**	1	Man-days of repair time per machine-shift*	0.34
Working days per year**	295		
Number of tractors required for working	35	Total number of repair man-days	3,536
Reserve machines needed*	9	Daily wage, roubles*	4.39
Total number of working tractors	44	Total wage bill	15,523
Coefficient of preparedness*	0.82	Total wage bill including regional coefficients and bonus payments	22,350
Total number of machines needed	54		
Actual number	54		
Shortage/excess	0		

Source; Rodigin, 1977a and b

By the end of September this suggested plan should have been worked out and after that it is sent to the combine if a four-tier structure is in force (see Chapter 3) or to the *obyedineniye*. Here all of the plans are collected and the *obedineniye* then coordinates and collates the figures, possibly adjusting them slightly, and passes the figures up to the next stage in the planning process, the Ministry. The Ministry collates the figures itself and, by this time, the detail in the plan has become considerably less than that which existed at the *lespromkhoz* level. Basic aggregate figures such as total production, total labour requirements, labour productivity, profitability and the main inputs such as fuel and equipment are what remain and it is these figures that the Ministry presents to *Gosplan*.

Gosplan adjusts the plans itself to ensure that the plans which, by October to November, it is receiving from all of the various ministries and other organs involved in the planning process, are consistent. This ensures, for example, that the output of fuel is sufficient to supply all of the requirements of the economy as well as those put forward by *Minlesbumprom* for chain saws, tractors, lorries etc. If it is not *Gosplan* will adjust the figures slightly so that the plans become consistent either by reducing the output of the various fuel-using concerns or by requiring an increase in the output by the fuel producers, or some mixture of these two strategies. Alternatively, if the investment and building plans need more timber than the timber producers have suggested then either the building plans will be reduced or the timber producers will have to produce more timber.

This process is often left incomplete and it is known that plans are sometimes inconsistent in the sense described above. By early December the Supreme Soviet of the Soviet Union meets to verify the most general directive indices in all of the plans. For some of the most important goods in the Soviet Union the plans go one stage higher to the Council of Ministers to be verified; this is the case with all types of timber. After this, the plans work their way back down the ladder, via Ministries, *obyedineniye*, combines if these exist, to the *lespromkhozi*. The main indicator figures should arrive at the *lespromkhoz* by the end of December before the year which the plan describes. Thus, the plan which the director of a *lespromkhoz* receives at the end of December should, subject to one or two alterations made as the plan went up to the Council of Ministers and down again, be the same plan as was submitted by him in late September (Omarov, 1974).

The Nature of the *Tekhpromfinplan*

The plan that a *lespromkhoz* receives is divided into ten sections. These are the basic plan indicators, the plan of production and sales, the supply plan, the plan for sylvicultural work, the plan for raising the efficiency of production, the investment plan, the labour and wages plan, the plan for costs, the plan for profit, profitability and the size of the economic incentive funds, and the financial plan. The first section of the plan, the basic plan indicators, contains usually thirty-one indicators of performance to be fulfilled during the year and these are listed in Table 7.2.

TABLE 7.2 Basic Plan Indicators

* per person

	Units
1. Timber sales	roubles
2. Commodity output	roubles
3. Total timber extracted	cubic metres
including usable wood, of which	cubic metres
(i) low quality wood for reworking	cubic metres
(ii) pulpwood	cubic metres
(iii) chips	cubic metres
4. Production of pit props	cubic metres
5. Sleepers for railways	units
6. Sawn timber	cubic metres
including export grade	cubic metres
7. Boxes and packing material	cubic metres
8. Split wood	cubic metres
9. Timber floated	cubic metres
10. Volume supplied directly to the final user	cubic metres
11. Resin produced	tonnes
12. Profit	roubles
13. Profitability	per cent
14. Payment to or receipt from the State budget	roubles
15. Remaining profit after budget contribution	roubles
16. Economic incentive funds, including	roubles
(i) the fund for material incentives	roubles
(ii) the fund for social-cultural improvement and construction of accommodation	roubles
(iii) the fund for the development of production	roubles
17. Work-force	persons
18. Wages fund	roubles
19. Average wages per person	roubles
including the growth in wages	per cent
20. Productivity of labour overall	roubles*
including the growth in productivity	per cent
21. Productivity of labour in logging	cubic metres*
including the growth in productivity	per cent
22. Cost of output	roubles
23. Expenditure on one cubic metre of output	kopecks
24. Reduction in costs	per cent
25. Cost per cubic metre	roubles
26. Productivity of capital per 1,000 roubles' worth of capital equipment	kopecks, cubic metres
27. Value of centralised investment	roubles
28. Changes due to centralised investment in	
(i) capital stock	roubles
(ii) productive capacity	cubic metres
29. Value of non-centralised investment	roubles
30. Changes due to non-centralised investment in	
(i) capital stock	roubles
(ii) productive capacity	cubic metres
31. Average value of the capital stock	roubles

The plan for production and sales. This is a very large part of the plan, and for *lespromkhozi* it is itself subdivided into twenty-five sections broken down on a quarterly basis with some of the contents further subdivided into individual plans for each *lesopunkt*. Most of the elements of this part of the plan are worked out in physical units for logging concerns, whilst for woodworking and furniture enterprises value units are most often used.

The supply plan. The supply section of the plan is broken down into quarterly figures. Usually the supply of fuel, lubricants, spare parts and also the larger items such as tractors, chain saws and other pieces of equipment are contained in this section. The State Supply Agency *Gossnab* deals with both the distribution of these inputs into the forest industry as well as most of the final distribution of timber and other products produced by the *lespromkhozi*.

The plan details the amount of the various supplies to which the concern receiving the plan is entitled (in the form of a maximum limit) as well as the payment required, the place where the supply will be received, the frequency with which the *lespromkhoz* may take delivery as well as whose transport will be used in delivering the supplies. The norms that are the basis of this part of the plan, for example the amount of fuel used by a chain saw per hour and so on, are worked out by the various research agencies in the Soviet Union. These norms can last, unaltered, for up to ten years before being re-assessed. About two years of tests are carried out on new machines to establish norms; the norms are usually different for different areas and different conditions.

The plan for sylvicultural work. This is a relatively small part of the plan and contains only quarterly figures for the various types of sylvicultural work to be undertaken. Usually this would be limited to brash clearance.

The plan for raising the efficiency of production. This is becoming a very important section of the plan as the requirements for increased labour productivity become more pressing. It can be seen from Table 7.2 that two types of efficiency are defined, overall productivity quoted in terms of roubles' worth of output per person employed in the *lespromkhoz* as a whole, and the more usual indicator of output in terms of cubic metres. In this latter case the work-force considered is not the overall number of people employed at the enterprise, the administrative personnel being excluded from the calculation. Also the effectiveness of the use of capital is presented in the plan here, the basic indicators in Table 7.2 contain a figure for the productivity of capital per 1,000 roubles' worth of capital equipment. However, this indicator is not as important as the figures relating to labour productivity.

The investment plan. This section contains only new investment and the labour requirements and wages needed to fulfil the planned investment, for example road building.

The labour and wages plan. This is another important section of the plan and is broken down into quarterly figures. It contains not only the figures for the wages to be paid (an important link with the financial system of the Soviet Union enabling some sort of control over the numbers employed) but also plans for the use of labour time.

The plan for costs. This is the largest part of the plan and runs to thirty-five subsections, many of them broken down into quarterly figures. The costs calculated vary from the overall cost of output to the cost of fuel. The emphasis is on cost reduction.

The plan for profits, profitability and for the size of the economic incentive funds. The emphasis on this section is more recent and the economic incentive funds were to play a large and important role in planning - see a later section for their discussion. Recently, however, they have become less signficant.

The financial plan. This section of the plan, like the wages plan, ties the productive concern into the financial system of the country. This allows a greater degree of overall control, one line of control coming from *Gosplan* via the ministries and concerning itself mainly with production and one financial line of control coming via the State banks and the financial plan.

Changes in the Planning Process

With the large number of figures that are contained in the 160 or so pages of a *tekhpromfinplan* there will be a number of major indicators whose fulfilment is critical and a number which are of little interest at all. An examination of this point is one that is often made in looking at the efficiency of the planning system in an overall sense. Obviously, if only a few of the indicators listed are of significance to the personnel in the logging concern in terms of the determination of their wages or bonuses there is likely to be some leeway in decision-making at the *lespromkhoz* level. If there is this latitude for individual 'interpretation' of the plan it follows that it is possible that the decisions made within the concern, whether by the director or by a feller on the forest floor deciding to fell one tree rather than another, may not necessarily reflect the wishes of the overall plan-makers, *Gosplan*. This point will be examined in the next section.

Also as, for the most part, the plan originates in the *lespromkhoz* itself and not with the Council of Ministers or *Gosplan*, there is no real reason why the suggested plan presented to the *obyedineniye* by the director and thence via the Ministry to *Gosplan* should bear any relationship to the true productive capacity of the enterprise in question. Indeed, if the rewards received by managers and workers depend upon the fulfilment of a plan, it is in their interests to submit a plan that is easy to fulfil, that is one which requires them to produce an amount which is less than their true capacity and hence relatively easy to produce. This point will be looked at in the second section.

The emergence of value units. Until the meeting of the Central Committee of the Communist Party of the Soviet Union in September of 1966, the financial indicators of the plan were not the major ones. The indicator that had to be fulfilled before all others was production and, as far as logging enterprises were concerned, this was expressed usually in physical units. The two important aspects of this indicator were 'production' and 'physical units'. The indicator of production meant that there was an incentive to produce units regardless of their use for the national economy. Those products that were easy to produce could be produced and stock-piled and the plan would still be fulfilled. The efficiency of the system depended upon the speed with which *Gosplan* could react to changes in patterns of demand. Also, the expression of output in terms of physical units meant that those trees most easily processed would be those actually chosen for processing. For example, as coniferous timber is easier and cheaper to fell and process than broadleaved species, it would be easier to fulfil a plan expressed in terms of 'cubic metres of output' if the output was in the form of coniferous species. Thus, in a mixed forest it would be better to fell, extract and process the coniferous timber and leave the broadleaved trees standing. This is what occurred and so, in the longer term, areas of forests were cleared of coniferous species and the other species were left standing. This meant that the seed trees left and the regrowth after logging an area over was the less desirable broadleaved species. The major financial indicator was that of costs and, whilst this was less important than output, the goal of cost minimisation tended to lead to a reduction in the quality of output.

Because such problems were general in the Soviet Union a reform was instituted in 1966 to improve the planning system. During this time there was a movement, led by the mathematical economists, to decentralise decision-making. It was hoped that, if the price system was improved, prices could be used as an indicator for individual decision-makers. For example, if the price of coniferous timber was set equal to its 'desirability' to the Soviet economy a logging enterprise could merely be told to maximise profits and coniferous species would be felled if the costs of doing so were less than the price received for the timber. Thus a system not dissimilar to a competitive market system was under consideration. The techniques involved methods similar to linear programming with the shadow values of the various inputs to the process under consideration being used as the price of that input for the enterprise producing it (Fedorenko, 1974). The discussion about such economic reforms have been covered in considerable detail elsewhere (Ellman, 1971 and 1973) and, whilst the reforms introduced in 1966 were by no means as far-ranging as the mathematical economists had hoped for, the thrust of the reform was to increase the use of value indicators and prices and to increase the freedom for decision-making at the enterprise level.

The main plan indicators after the reform were value of sales, volume of timber extracted, percentage production of useful timber, the productivity of labour, profitability, the total wages fund, costs and the economic incentive funds. These were all listed in Table 7.2 earlier. In general, value indicators were made more important and profit became an important part of the plan. This synthetic indicator meant that there would be no incentive to reduce costs if this were associated with a loss of quality in the output, as lower quality output would have a lower price and so would act to reduce profits and profitability. The production indicator was dropped in favour of 'sales' to reduce the incentive to stock-pile unwanted output to fulfil the plan.

Clearly, for value indicators to be of any usefulness, the prices of the various outputs and inputs must be 'correct' in the sense that the most desirable outputs have the highest prices and the scarcest of inputs to the productive process must be the most expensive. In that case, if the profit criterion is used, the most desirable output will be produced with the minimum expenditure of scarce resources. The reforms, therefore, coincided with a change in the prices for timber assortments. The next half of this chapter will discuss this. As far as forestry is concerned, however, several important results emerge from the reforms.

First, for the logging industry directly, the plans which were contained in the *techpromfinplan* but which applied to the individual *lesopunkts* within the enterprise remained virtually unaltered. All of the indicators related to the activities of felling, transporting and woodworking at the lower landing are described in physical units. So on the forest floor the same problem remains. As coniferous trees are easier to fell, transport and to work on at the lower landing, broadleaved trees are often left standing as an area is cleared of its coniferous timber. Indeed, up to 40 to 50 per cent of the growing stock on a particular site might be left standing for this reason. The effect that this has on the productivity of labour when the new machines are used has already been noted in Chapter 5. This point will be reinforced by the fact that the director of a *lespromkhoz* will rarely find it in his interests to insist on true clear-felling as the price of broadleaved timber is less than that for coniferous timber and so, on a cubic metre basis, broadleaved timber is less efficient at fulfilling the profitability requirements of the overall plan. Also the price of chips produced from low quality wood has, in many cases, been set too low to make it worthwhile to extract low quality wood for conversion into chips at the upper or lower landing; rather it is left on the forest floor.

A second factor which arises from the use of the value indicators and profitability in the plan affects forestry indirectly. The factories producing the capital equipment required to make forestry more efficient are operating under the directions of their own plan with its insistence on the value of output and profitability and its incentives to reduce costs. But whilst new machines, for example grapple skidders or feller-bunchers which would increase labour productivity in logging, have higher prices and so will increase the value of a tractor factory's output, they have significantly higher costs of production in the initial stages of their production. So the production of radically new machines will lead to the fulfilment of one indicator, provided enough are produced, but to the underfulfilment of the profitability and cost indicators. This has led many such producers to 'play safe' by producing either the old equipment or to produce 'new' equipment which varies very little from the old. Thus, in the case of tractor production, new tractors are produced which are only slightly revamped versions of the old ones - larger engines maybe. This policy keeps the costs of innovation down to a minimum and keeps profitability as high as possible whilst still producing 'new' machines. But it does not lead to the production of the new types of machines needed by the logging industry. This point will be looked at in more detail in the next chapter.

Another problem caused by the plans that are presented to the tractor-producing factories has been discussed in Chapter 5. The fulfilment of the plan in terms of the number of spare parts produced has a very low priority for the producers of capital equipment compared with the production of the basic capital equipment itself. This has led to shortages of spares and an increase in machine downtime. Some experts argue for a doubling of the production of spare parts. The effect of the lack of spares is shown by the differences between the 'coefficient of preparedness' of tractors in actuality and the norms used in planning. The norm in planning for the TDT-55 is 0.82 as shown in Table 7.1; that is to say the tractor is expected to be operational for 82 per cent of the time. In reality the figure is less than this, nearer to 75 per cent and for the newer machines it is even less (Moshonkin, 1975).

Soviet authors tend to blame this shortage on the fact that spare parts are not a major part of the plan and are planned in physical units - see Chapter 5. But it is likely that there is a more deep-seated reason. The whole idea of reliability and spare part production (a very important aspect of production in an increasingly technological environment) implies a recognition of random and uncertain factors in the operation of capital equipment and so the influence of risk and uncertainty on production. In the West any course of, say, managerial economics would contain the discussion of decision-making tools to be used in conditions of risk and uncertainty which would be applicable to the idea of the optimum level of spare parts inventories and production. In the Soviet Union there is little emphasis in training managers in the skills of decision-making under risk and uncertainty; indeed it could be argued that the incorporation of risk at this level is inappropriate given the Soviet Union's method of planning. The planning system is based upon a parallel system of responsibility - the planned targets have to be someone's ultimate responsibility or the planning system would break down. Clearly the introduction of ideas of uncertainty into decision-making suggests that factors other than the input of the managers or workers affect output and so would weaken the link between output and the performance of the staff and workers of a *lespromkhoz*. Thus, a poor performance could be blamed on outside uncertain forces and rewards could not necessarily be claimed for high levels of output. Given the inadmissibility of this position in the Soviet Union, it follows that the ideas of risk and uncertainty are not going to be central to decision-making in the country and this will be reflected in a 'blind spot' in the production and planning of spare parts. This might not be the case if the idea of the use of new technology was tackled by someone in the West with a good grounding in decision models incorporating risk and uncertainty.

The nature of norms used in planning is also likely to be incompatible with the 'stochastic' nature of unreliability. If a tractor requires say, S spare parts per annum and a *lespromkhoz* has T tractors then the *lespromkhoz* should receive S.T spares as part of the plan and the total number of spares to be produced would be S.N if N were the total number of tractors in use. However, as unreliability is stochastic in nature, some *lespromkhozi* will require more spare parts, and so will experience extra machine down-time and some will need less and so be able to keep their machines in use*. Thus, a system of planning based on simple norms is unlikely to be able to cope with this problem of spares without some over-production and a dissemination of extra parts to act as an insurance for the possibility of machines being less reliable than average. What is needed in this situation is a system of planning and distribution that is very responsive to 'consumer' demands, so that those *lespromkhozi* which need spares can get them quickly. Whilst this is achieved in the Soviet Union to a limited extent, partly officially and partly unofficially (a recent article in the newspaper *Lesnaya Promishlennost* told how one fitter could get the spare parts required only if he were given a bottle of vodka or two), it requires in the future a consumer oriented system of planning that does not square with the centralised control that is embodied in the economic system of the country.

The building of forest roads is planned in a similar way for *lespromkhozi* as spare parts are for the tractor producers. It forms a very insignificant part of the plan. As can be seen in Table 7.2 it is contained in the value of non-centralised investment. It is not something that rewards and bonuses depend upon. Thus the fulfilment of the roading plans is not pressing and the effects of this have been seen in Chapter 3. For specialised construction concerns such as *Soyuzlesstroi* the plans used to contain indicators in terms of building starts and money spent on construction. This has been the case until now and means that there was no incentive to complete buildings and construction work. As a result, from 1976 to 1979, the actual new *lespromkhozi* built in the logging industry were capable of an annual throughput of 19.6 million cubic metres compared to a planned amount of 26.1 million (Batrakov, 1980). In the previous chapter the under-fulfilment of the plans for the forestry workers' accommodation was mentioned. Soon the plans for construction are likely to be changed in favour of indicators such as 'work completed' in order to avoid this very problem (quoted in Nove, 1980).

The problem of taut planning. The fact that the basis of the *tekhpromfinplan* is worked out by the director of a logging concern who will, along with his staff, receive rewards in the form of monetary bonuses and non-monetary privileges if the plan is fulfilled, but will not if it is under-fulfilled, means that there is an incentive for the director and his planning department to conceal the true productive capacity of their *lespromkhoz*. For example, if the *lespromkhoz* could process 100,000 cubic metres of timber per annum at full capacity, it would obviously not be in the director's best

*This is similar to simple queuing theory where the arrival rate of tractors needing spare parts is equal to the service rate, the number of spare parts. This leads to queues of infinite length.

interests to formulate a plan which suggested an output of 100,000 cubic metres in the next year. First, there would be no room for the vagaries of the climate or for the suppliers who have to supply inputs for the logging concern, fuel, spare parts, etc. Thus, an underestimation of capacity may be seen as a form of risk aversion on the part of the planners in the various enterprises. Secondly, as the plan goes to *Gosplan* and comes back again, it is possible that it will have been altered slightly so that the director and his planning department receive, at the end of December, a list of indicators from their *obyedineniye* requiring that they produce 110,000 cubic metres of timber in the following year. Therefore, a problem which the Soviet planners recognise is that the plans which are formulated may not be 'taut', that is to say, they may contain elements of slack which have been inbuilt into them for the reasons just described.

To overcome this problem three funds were set up under the umbrella term of economic incentive funds. The first of these is the fund for material incentives (FMI) which is used to pay various bonuses and prizes. The use of the fund for such payments is agreed with the union representatives. Usually about half of the FMI goes to pay bonuses for actual production and these are paid out at the end of the year. They are often referred to as the 'thirteenth month's pay' for this reason and, because they are only paid to those workers who are still at the *lespromkhoz* at the end of the year, they act as some incentive to reduce labour turnover. This use of the FMI represents bonuses paid for quantitative performance, volume felled during the year, etc. The rest of the FMI goes in the form of prizes for qualitative performance, for example a high level of labour productivity in a brigade or a low breakdown rate for the equipment used, etc.

A second fund is the fund for social-cultural improvement and construction of accommodation (FSCICA). This is used mainly for the construction of living accommodation, often with better quarters being built for the better workers so that this will reinforce the FMI in giving an incentive to workers to perform well. Some of the fund goes for the building of nurseries for the workers' children, for holiday 'camps' for the workers' use and for other such purposes.

The third fund set up was the fund for the development of production, the FDP. This was a fund which could be used for investment within the *lespromkhoz* or as a basis for credit from the State bank for investment. In a sense the FDP resembles retained earnings but can only be used within the concern generating the fund. There is no inter-*lespromkhoz* financing, so surplus units cannot fund deficit ones via a money market of some type. This aspect of finance is taken care of by the payment of money or receipt of cash from the State budget - see element 14 in Table 7.2. Surplus units, that is *lespromkhozi* which for reasons of, say, location, find it easy to produce valuable timber at low cost have to pay a sum into the State budget before calculating their operating profitability whilst those that operate under more adverse conditions receive a subsidy. In the early 1970s the FDP was used for the modernisation of equipment and the introduction of new technology in basic operations (40%) for the modernisation of transport (36%) placed in the bank as deposits (22%) and the remaining two per cent was used for other uses (Glotov *et al.*, 1975).

The first two funds, the FMI and the FSCICA, are those which are supposed to have most effect on ensuring the formulation of taut plans. The values of these funds are set in the five-year plan, usually the FMI is equal to about 10 per cent of the wages fund and the FSCICA is equal to 30 per cent of the FMI. However, the value of these funds and hence the benefits that flow from them can be manipulated by the planners in the logging concerns. The funds are increased in size if the suggested plan submitted by the logging concern proposes a greater level of production or a higher productivity of labour than the figures in the five-year plan. Also, they are increased if the final agreed *tekhpromfinplan* is over-fulfilled.

To ensure taut plans the following system is used. Four 'fund forming indicators' are chosen. For forestry and logging these are the value of timber sold, the production of useful timber excluding firewood in volume terms, the productivity of labour and profitability. An increase in these indicators in the suggested plan over the figures for the relevant year in the five-year plan will increase the values of the FMI and FSCICA if the plan is fulfilled. The amounts by which the funds increase for a 1 per cent change in the fund forming indicators are shown in Table 7.3. As can be seen, if the suggested plan is drawn up by the director of a logging concern with his planning department which incorporates a 1 per cent increase in the value of sales over the figures in the five-year plan, along with a 1 per cent increase in the production of useful timber, a 2 per cent increase in the productivity of labour and no change in overall profitability this would lead to an increase in the size of the funds of 3.1 per cent if the plan were exactly fulfilled in terms of these indicators.

Extra rewards are incorporated for over-fulfilment. These norms are also shown in Table 7.3; they are 0.7 times the norms just outlined. Thus, if the plan described above was suggested and accepted and at the end of the year the situation was that, compared to the figures in the five-year plan, a 2 per cent increase in the value of sales had been realised along with a 1 per cent increase in the production of useful timber, a 2 per cent increase in the productivity of labour and a 2 per cent increase in profitability, the value of the funds would increase by 4.78 per cent. However, it is clear from the formulation of the coefficients that it would have been better to express fully the potential of the logging concern in the suggested plan. If the figures just referred to had been incorporated in the plan and had been fulfilled as suggested, the funds would have increased by 5.5 per cent.

Thus, the use of such norms should lead to taut plans. To ensure that the suggested plan are realistic, there are penalties for under-fulfilment which are 1.3 times the basic norms. These are also shown in the table. Thus, if a plan is suggested that contains figures for the productivity of labour 1 per cent above the figure in the five-year plan, but in the end the productivity of labour is the same as that found in the five-year plan, the funds would be reduced by 0.39 per cent.

The effectiveness of such a system in ensuring taut plans is very difficult to measure. When the system was introduced the

TABLE 7.3 The Fund Forming Indicators

Indicator	Change in the funds for a 1 per cent increase in the figures in the five-year plan	Increase for every extra 1 per cent achieved over the suggested plan	Penalties for under-fulfilment
Value of sales	2.0	1.40	2.60
Production of useful timber	0.5	0.35	0.65
Productivity of labour	0.3	0.21	0.39
Profitability	0.2	0.14	0.26

Source: conversations with forest economists at the Kirov Forestry Academy

enterprises chosen in the early days to test the system showed great increases in production. Sales increased by 24 per cent in some cases and profitability by up to 78 per cent (Moskin, 1968). The *lespromkhozi* and *obyedineniye* now often fail to fulfil their plans or only just manage to fulfil them which might be explained by very taut planning.

Paradoxically the system has doubtless contributed to the re-emergence of the volume indicators as the main effective plan indicator. The system is such that the four indicators chosen are the main ones to be fulfilled and concentrated upon and the coefficients are such that value of sales is the most important one - especially as a 1 per cent change in the amount of timber harvested is probably easier to achieve than the changes in the other indicators. To increase the funds by the same amount as a 1 per cent increase in output would require a 4 per cent increase in the output of useful timber, a 6.7 per cent increase in the productivity of labour and a 10 per cent increase in the productivity of labour. The low price of broadleaved species means that an increase in the value of sales is best achieved by felling coniferous timber rather than totally clearing the site if the growing stock contains both coniferous and broadleaved species.

In fact, in terms of publicised performances of *lespromkhozi*, the value of sales is dropped and volume figures are universally quoted. This means that the effect of the value indicator which would put a premium on the production of better quality and therefore higher priced timber has been lost. The publicised figures form part of the working of the planning system by exerting some sort of moral coercion for plan fulfilment. *Lespromkhozi* that do not fulfil their plan, usually in terms of volume produced, less often in terms of productivity and very rarely in terms of other indicators, are named in the press along with their management. Workers and managers who do well are interviewed, have their photographs published sometimes on the covers of the technical journals and are given medals and awards. These will entitle them to preferential treatment, from being served first in the queues so common in Soviet life to extra pay and better living accommodation.

As well as the volume produced, the productivity of labour has become very important; the reasons for this are partly the fact that it is one of the fund forming indicators and partly because of the increasing emphasis being placed on intensive growth in the industry. However, this does not mean that the value indicators are of no importance. The price system will be discussed in the next section.

Internal planning. The emphasis on planning means that many internal plans are drawn up for the fulfilment of the staff within an enterprise. Mainly monthly or half-monthly plans are produced for each *lesopunkt* or brigade etc., and these are consistent of course with the overall one-year plan for the enterprise. The balance method of planning is used in the formulation of these plans, so that the requirements for one part of the *lespromkhoz* equates with the resources that are available. They are similar to much scaled down *tekhpromfinplani* with sections on production, usage of capital equipment, pay and the wage bill and the productivity of labour.

These plans are either on a monthly or half-monthly basis and the choice depends upon the conditions in an area. Half-monthly plans are usually used in the heavily forested areas of the country. They are designed to produce a more even flow of production and to maintain a level of incentive by making the planning period relatively short. Thus, bonuses paid for the fulfilment of these short-term plans are paid every fortnight and are sufficiently close to receipt at all times, even the beginning of the fortnight, to influence work levels. The monthly plans minimise the funds required to pay bonuses, as a good showing in the first half of a month, which would result in a payment of a bonus if the half-monthly plans were in force, might go unrewarded if the second half of the month is not so successful. Such a method of planning is often used in the lightly forested areas and for preparatory and auxiliary operations.

Every now and then one-hour plans are experimented with. In the 1950s Koroleff (1952) reports their use. Large notice-boards were set up at the side of the logging area with the brigade's performance displayed each hour and the required output shown beside it. The plans, of course, need much care in their drawing up and the conditions and abilities of the various brigades have to be considered. In 1976 one-hour plans were experimented with again; in both cases, however, they were dropped as being unwieldy. The internal planning system is similar to the external system in that workers can 'volunteer' for extra output over and above the norm to be included in their plans for extra bonuses.

PRICE FORMATION IN SOVIET FORESTRY

Introduction

The institution of the value of sales and profitability as two of the major indicators in the plan will only ensure optimum decision-making at the level of the *lespromkhoz* if the species which are the most desirable have the highest prices for the logging concerns and the resources needed for their production are priced according to their scarcity. This meant that the introduction of the reforms in the planning system required a price system which reflected the true desirability of the timber assortments and the resources in the country. Thus, the reforms in the planning system in 1966 were coupled with a wide reform of the prices in the Soviet Union. Before looking at the success of the price reform and the influence that this has on the logging enterprises, it is necessary to look at the price system in general in forestry in the Soviet Union.

Price Formation in the Soviet Union

There are really two types of prices in the Soviet Union. One is the set of prices formed in the markets for agricultural products in most of the towns in which farmers from the *kolkhozi* sell their produce from the private plots; here prices are more-or-less free to fluctuate according to the laws of supply and demand. From the point of view of forestry, however, the only important prices to consider are the second sort of prices, the prices that are determined by the State. A committee, part of *Gosplan*, has the duty of setting prices for all of the products that are produced in the Soviet economy. This is no small task. Recently prices were reformed but until 1982 the price list for logs and other products of logging concerns (*Preiskurant No 07-03* (Anon, 1967)) had remained in force more-or-less unchanged from 1967. In this section the examples are taken from this price list and the new price list that has just come into force will be discussed briefly later.

Prices in the Soviet Union are calculated on the basis of costs. There are three methods of price determination used in the formulation of prices for different types of commodities and the method employed for forest products is the most common of the three. In forestry the price of timber is calculated on the basis of the average cost of production on a country-wide basis. The costs included are those associated with raw materials, intermediate inputs such as fuel and spare parts, wages for all those involved in production, the charges for national insurance which usually represent about 4.5 per cent of the wage bill, a stumpage tax paid to the State and the depreciation charges from the use of capital equipment. Expressed on a per cubic metre basis this will give the average cost of production for timber at the lower landing (Kozhin *et al.*, 1975).

This is used as the basis for the price of timber assortments that the *lespromkhoz* receives on sale of the timber. The price P is given in the equation below. Here AC is the average cost of timber

$$P = AC \times (1 + p) \times RK \times QK$$

production in the country as a whole, p is the rate of profit for the industry, RK is a regional coefficient and QK is a quality factor.

The quality factors, QK, vary according to the end use of the timber. For saw logs for general use the quality factor is equal to unity. For fuelwood the coefficient is around 0.15 to 0.2 whilst for wood suitable for very high quality use, such as the manufacture of musical instruments it is 2.5.

The regional coefficients, RK, are an attempt to differentiate the price of timber throughout the country so that there will exist an economic incentive to use timber from the areas where there are abundant supplies and to economise on the use of timber from the lightly forested parts of the country. The regional factors are determined so that they are lower in the area of abundant timber stocks and higher in the lightly forested areas. This leads to a distribution of prices such as that shown in Table 7.4. In this table it is evident that *lespromkhozi* in the Irkutsk area will only receive 8.6 roubles per cubic metre of coniferous timber of the grade in question but logging concerns in the Ukraine will get 23.3. This is designed to encourage the organs distributing timber to use timber from the cheaper regions which are those with a greater timber stock. *Gossnab*, which is the organisation which will distribute most of the timber felled by *Minlesbumprom*, is profit oriented and so will wish to purchase timber for distribution in the cheapest areas. In this case, this will be from the Irkutsk area.

TABLE 7.4 Prices of Round Wood Suitable for Sawn Timber for Construction and General Use

	Price per cubic metre for pine and spruce *	Price per cubic metre for birch **
Komi ASSR	13.4	12.8
Karelia	11.2	7.1
Leningrad area	12.9	7.1
Irkutsk area	8.6	5.5
The Ukraine	21.3	18.7

* Quality grade II, length of bolts from 4.0 to 6.5 metres and 14 to 24 centimetre top diameter.

** Quality grade II, length of bolts from 3.0 to 5.0 metres and top diameter in excess of 14 centimetres.

Source: Anon, 1967

These are the prices which the logging concerns receive, the price f.o.b. or *optovaya tsena predpriatia*. The price which the sawmills and other timber users have to pay, the price c.i.f. or *optovaya tsena promishlennosti*, is calculated by adding a certain amount onto the price f.o.b. It might be thought that the process would be a simple matter of adding the transport cost onto the price f.o.b. to arrive at the price c.i.f.. However, in the system of prices for Soviet timber the method is a little more complex; a system of additions or rebates, *nabavki* and *skidki*, are applied to the price f.o.b. in an attempt to make the planning system more efficient.

For the purposes of determining the price c.i.f. that the timber users are required to pay the country is divided into five areas or regions. These are not necessarily contiguous regions but rather are groups of areas with similar degrees of overall forest cover. Areas of the country that are heavily forested such as the Komi ASSR, the Siberian north and the Far East are classified as region I. The area around Archangelsk and Murmansk are in region II and areas like the Ukraine, Georgia and other parts of the southern Soviet Union are classified as region V. The additions to the prices f.o.b. that are applied to coniferous and broadleaved timber originating in the Komi ASSR are shown in Table 7.5. The additions are dependent upon the destination of the timber. Thus the addition to the price of a cubic metre of coniferous timber destined for use in region I is 4.30 roubles, giving a price c.i.f. that the sawmill, say, will have to pay of 17.7 roubles per cubic metre for the type of timber referred to in Table 7.4 - that is 13.4 roubles plus 4.30 roubles. If, however, the destination of the timber was a sawmill in the Ukraine in region V the price per cubic metre would be 27.2 roubles. The prices of coniferous round wood and birch round wood paid by the timber using concerns are shown in Tables 7.6 and 7.7.

TABLE 7.5 Additions to the Price f.o.b. for Timber from the Komi ASSR to Arrive at the Price c.i.f.

	Region of Recipient				
	I	II	III	IV	V
Coniferous round wood	4.30	5.00	6.50	9.50	13.80
Broadleaved round wood	1.00	1.40	2.30	4.40	7.60

TABLE 7.6 Prices of Logs of Coniferous Species Suitable for the Manufacture of Sawn Timber for Construction

Quality grade	Top diameter	Region of Recipient				
		I	II	III	IV	V
I	26+	21.50	22.20	23.70	26.70	31.00
II	26+	18.90	19.60	21.10	24.10	28.40
	14-24*	17.70	18.40	19.90	22.90	27.20
III	26+	17.50	18.20	19.70	22.70	27.00
	14-24	16.30	17.00	18.50	21.50	25.80
IV	14+	12.60	13.30	14.80	17.80	22.10

The figures all refer to the price c.i.f. per cubic metre of timber of spruce, pine, fir or cedar, in the form of logs of length between 4.0 and 6.50 metres. The row marked with the asterisk refers to the quality of coniferous timber referred to in Table 7.4 and the prices in that row are found by the addition of the relevant factors from Table 7.5.

TABLE 7.7 Prices of Birch Logs Suitable for the Manufacture of Sawn Timber for Construction and General Use

Quality Grade	Top diameter	Region of Recipient I	II	III	IV	V
I	14+	15.30	15.70	16.60	18.70	21.90
II*	14+	13.00	13.40	14.30	16.40	19.60
III	14+	11.30	11.70	12.60	14.70	17.90
IV	14+	7.80	8.20	9.10	11.20	14.40

The figures refer to the price c.i.f. per cubic metre of birch in the form of logs of length between 3.0 and 5.0 metres. The row marked with the asterisk refers to the quality of birch timber referred to in Table 7.4 and the prices in that row are found by the addition of the relevant factors from Table 7.5.

Both of these tables are taken from Anon, 1967.

It should be noted that all timber users in a given region will pay the same price for timber of a certain grade regardless of its place of origin. The factors added to the price f.o.b. differ according to the region of origin. In Table 7.5 these factors were shown for the Komi ASSR. For grade II round wood from the Komi ASSR going to region V the price f.o.b. of 13.4 roubles per cubic metre was increased by 13.8 roubles to give a price c.i.f. for a sawmill in the Ukraine, say, of 27.2 roubles. If the timber had originated from Irkutsk, say, the price f.o.b. of 8.6 roubles (see Table 7.4) would have been increased by a factor of 18.6 roubles to give the price c.i.f. in region V of 27.2 roubles.

The purpose behind the formation of these five regions with their different *skidki* and *nabavki* is an attempt to concentrate production in areas with relatively abundant timber supplies and to reduce the level of production of timber products in lightly forested regions. The rationale behind this is that it is cheaper to process timber in the areas where timber grows and then to transport the processed products to the areas in the populated parts of the country, mainly central European parts of the RSFSR, than it is to transport bulky round timber to processing plants in the populated parts of the country. The price of timber in the lightly forested areas that timber users have to pay is much higher than in the heavily forested regions. This would decrease the profitability of production of timber products in the lightly forested areas (regions IV and V) and increase it in the heavily forested areas (region I).

Looked at from the point of view of the organs distributing the timber from forest to final user, it is possible that such a system leads to incorrect timber allocation. *Gossnab*, which derives part of its profit from the differences in the buying and selling prices of timber, will have an incentive to take timber from a low price area, such as *Irkutslesprom*, and supply it to a high price user such as a sawmill in the Ukraine. This would involve paying the *lespromkhoz* 8.6 roubles per cubic metre for grade II round wood (see Table 7.4) and selling it at 27.2 roubles (see Table 7.6). The optimum distribution strategy will clearly depend upon the actual costs of transporting timber, but if they are not carefully worked out it would pay the distribution agencies to transport timber needlessly long distances. This may be part of the explanation for the transport of timber over long distances that was mentioned in the latter part of Chapter 2.

For the sake of completeness, the range of prices determined in the Soviet Union from the raw material until the final product for consumers should be outlined. The price c.i.f. that a timber user has to pay for raw materials enters into the costs of production and, after an addition of the other costs and the relevant profit margin, becomes the price f.o.b. or *optovaya tsena predpriatia* for the finished product, a piece of furniture, say. After the addition of the transport costs required to get the finished product to the point of sale, the price c.i.f. or *optovaya tsena promishlennosti* for the object is obtained. The actual price paid by the consumer is this price plus selling costs, the profits of the organisation selling the product and a turnover tax which goes to the State budget. This range of prices for an item of furniture is shown in Figure 7.1. As far as we are concerned, only the first two layers of prices, (1) and (2), are of significance.

The Price of Reform of 1967

The price reform of 1967 was part of the overall reforms in the planning process. Prices in general were increased so as to raise the overall profitability of the logging enterprises. This had been necessitated by the increasing costs of timber harvesting caused by the movement of logging operations further east and north. The average price increase was 26 per cent and this had the immediate effect of increasing the profitability of logging (expressed as a percentage of the capital employed) from 3.7 per cent in 1965 to 20 per cent (Rodigin *et al.*, 1972).

However, the price reform had a more important aim than just increasing the loggers' profits. It was designed as a measure to ensure that, when presented with a *tekhpromfinplan* with objectives expressed in monetary units such as value of sales, etc., the individual *lespromkhozi* would react in a way that was in line with the overall planners' desires. For example, the desire of the planners is for a more complete utilisation of the timber growing in a particular area. This means that not only do the planners wish that the coniferous timber should be felled, but also the broadleaved species should be harvested and not left standing after logging operations. To this end they had to resolve the apparent conflict between making it profitable to fell the broadleaved trees, so meaning that the price f.o.b. had to be sufficiently high, and making it profitable for the users of timber to use the broadleaved species in their product, so meaning that the price c.i.f. had to be as low as possible. This was achieved, or at least attempted, as follows. Referring to Table 7.4, the difference between the prices of coniferous timber and broadleaved timber at the logging enterprise is relatively small. In the Komi ASSR, for example, the price is only

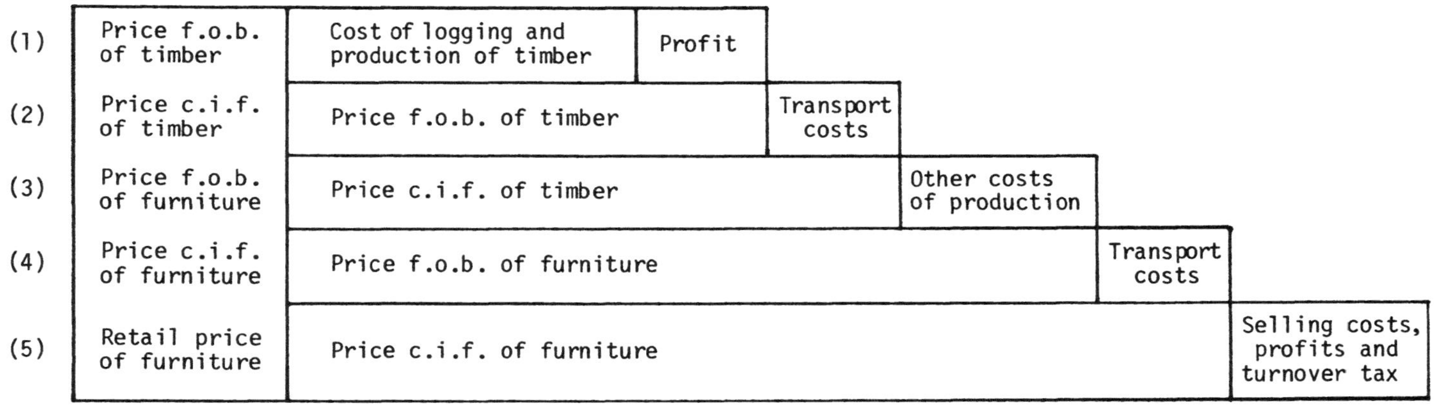

(1) - *optovaya tsena predpriatia*
(2) - *optovaya tsena promishlennosti*
(3) - *optovaya tsena predpriatia*
(4) - *optovaya tsena promishlennosti*
(5) - *roznichnaya tsena*

Source: Petrov, 1971b

FIGURE 7.1 The Prices Generated by the Manufacture of a Piece of Furniture

60 kopecks less for birch logs compared to coniferous logs of a similar quality. However, the prices c.i.f. that the timber users must pay show a wider divergence. Thus, for the same quality wood referred to in Table 7.4, in the case of coniferous timber that is quality grade II, with top diameter between 14 and 24 centimetres and for birch timber that is grade II with top diameter more than 14 centimetres, a timber user pays 4.7 roubles less in region I for birch, 5 roubles less in region II, 5.6 roubles less in region III and so on - see Tables 7.6 and 7.7. This is because the additions to the price f.o.b. to arrive at the price c.i.f. are differentiated according to broad species type - see Table 7.5. These factors are a lot greater for a cubic metre of coniferous timber than for broadleaved species.

On average, for all grades of timber, the prices f.o.b. for broadleaved species' logs were set 15 per cent lower than the prices of comparable timber of coniferous species, whilst the prices c.i.f. were about 30 per cent less (Kozhin *et al.*, 1975 and Bursin, 1978). This was achieved by differences in the additions made to the prices f.o.b. so that 40 per cent of the cost of transporting broadleaved timber was to be borne by the prices of coniferous timber. It is now clear, however, that this aspect of the reform has been unsuccessful. In the European parts of the country the allowable cut of broadleaved timber is only 50 to 60 per cent exploited - see Chapter 3. In many *lespromkhozi* the broadleaved species are left standing as the coniferous species are harvested. To fell a certain value or volume of timber the choice exists between felling all of the timber on a site, both coniferous and broadleaved, or felling only the coniferous trees on the specific site and also on adjoining areas in order to make up for the volume of timber left standing. The second course of action will necessitate an increase in the scale of operations in the extensive sense. The value of sales will be higher as the value of coniferous timber to the *lespromkhoz* is greater. Costs, however, might rise so reducing profitability for, although the basic production costs associated with coniferous timber are less, there will be a need for slightly increased transport costs as the area of forest logged over increases.

Overall, however, it would seem that the present economics of production favour the leaving of broadleaved timber unharvested and so are effectively precipitating the need to move to ever more remote areas to harvest timber. As the degree of mechanisation increases it is probable that the average costs associated with harvesting broadleaved timber will rise and the productivity of labour will fall in relation to the costs and productivity of harvesting coniferous timber. If the average tree size of broadleaved timber is somewhat less than that of coniferous trees, the costs of felling, transporting, delimbing and cross-cutting expressed on a cubic metre basis will obviously rise. The use of delimbing machines also will be less advantageous when applied to broadleaved species. The form of these is such that mechanisation of delimbing is less easy than the comparatively easy operation of delimbing a conifer with a well defined straight trunk.

The effects of the price differential on various species of timber is reinforced by another factor in the price structure - the price-size gradient. From Tables 7.6 and 7.7 it is evident that the price per cubic metre of small timber is less than that of larger sized logs. The prices in these tables are prices c.i.f., the prices f.o.b. of such timber will exhibit a similar absolute differential. As smaller timber, no matter what its species, is more expensive on a per cubic metre basis to fell, etc., it is again rational to incur slightly greater transport costs in harvesting only the largest timber from a larger area of forest rather than fell all of the timber on a specific site. Also, from the point of view of the timber users, it is profitable to pay more for the bigger timber and spend less on its conversion than to pay less for the smaller assortments and incur greater costs in production of the end products. As a result some suggested a greater differential for the prices c.i.f. of large and small timber - up to 20 to 25 per cent compared with the present 6 to 8 per cent (Bursin, 1977 and 1978). If this were coupled with a reduction in the differential in the prices at the *lespromkhoz* level by a range of factors such as those shown in Table 7.5 which were also differentiated in terms of timber size, the desired effects would be achieved. *Lespromkhozi* would have an incentive to harvest smaller sized timber of all species and the timber users would have an increased demand for it.

Another area where the price system was causing some distortion was in the pricing of low quality wood. In order to encourage the production of chips for the use of the particle board and pulp and paper makers, a new price category was created in the 1967 price list. This was the category of 'low quality wood for technological needs' which contained low quality logs, branches, etc., that were good enough for the production of chips rather than going into the lower quality category of firewood. In order to stimulate the production of chips and the more complete use of timber logged from an area, the price c.i.f. of such 'low quality wood for technological needs' was set twice as high as the price for firewood. However, once the transport element was taken into account, the price f.o.b. at the *lespromkhoz* of such wood was found to be greater than some low quality timber assortments such as wood for boxes and packing material. This was specifically the case in the Irkutsk, Krasnoyarsk and Amur regions. Thus, it became rational for *lespromkhozi* in such areas to under-fulfil their plans in terms of the production of boxes and packing material and divert the timber that would have been included in such categories to the newly created 'low quality wood for technological needs', so increasing the value of goods sold and profitability indicators (Kozhin *et al.*, 1975).

On the other hand, in many areas the price f.o.b. of chips was less than that required to cover the costs of production, especially if production was to be carried out at the upper landing or by mobile chippers operating on the forest floor (Bursin, 1978; Kozhin, 1972 and Sinyakevich, 1979). The profitability of chip production in general for *Minlesbumprom* has been 13.8 per cent but for their production in a *lespromkhoz* at a lower landing, as opposed to their manufacture at sawmills or specific pulp mills or board manufacturing plants, it is only 2 per cent. This has not created the necessary stimulus for the production of chips at the *lespromkhoz* level. To eliminate this problem the price of chips for paper production was increased by 23.7 per cent in 1978 so as to increase the profitability of chip production at the lower landing to between 19 and 20 per cent.

This, it is expected, will increase output by 600,000 cubic metres per annum (Bursin, 1978). Also, it is suggested by some authors that the price of off-cuts and waste from sawmills is too low to make it worthwhile to offer waste for working into chips at other enterprises.

The Recent Reform and Future Reforms of the System

Since 1967 there has been a general increase in the costs of production, as there was in the period before the 1967 reform, and this has had its effect on the profitability. Many more logging concerns were loss-making than was the case immediately after the reform. The price list of 1967, because it was so out-of-date, was in part a cause of some of the problems in Soviet forestry. It mitigated against the clear-felling of an area and so reduced productivity of labour and increased the extensive expansion of logging concerns. As a result, in 1982, a new price list was introduced (Romanov, 1982), in which timber prices were increased by 40 per cent on average to give a return on capital of 16.2 per cent. Prices have also been adjusted to encourage the harvesting of small-size timber. The effectiveness of this reform has yet to be judged.

One of the problems always associated with reform is an ideological one. The price of timber should be equal to 'the socially necessary labour input into the production of the good' - in line with the underlying theme of the labour theory of value associated with Marxist theory. However, in practice, it is very difficult to distinguish between those costs of, say, felling that should be incorporated into the price of saw logs or into the price of chips, both of which may come from the same tree. So there exists some leeway to arrange prices at levels not immediately equal to the costs of production including profits. Some texts take as their starting point the work of the mathematical economists whose view of the economy is one of an optimally functioning system based on concepts of mathematical programming. Here, the prices of goods are equal to their dual values in a linear programming setting so that their price represents the true value of the output that would accrue to the economy if one extra unit of timber was produced (Anon, 1968 and Glotov, 1973). Thus, the price of one cubic metre of timber should be equal to the net increase in output to the economy if one extra cubic metre of timber were produced - its marginal revenue product in neoclassical terms.

> 'In practical terms, the price of logging and sawmill wastes should be set at the size of the economic effect that would be achieved with their use as a raw material'.
>
> Lobovikov and Petrov, 1976

In this quotation, Lobovikov and Petrov show that they accept and advocate such an approach to pricing in the limited area of wastes from logging and sawmilling. However, this programming approach to the economy seems to be falling from favour which is unfortunate as it offers some means of breaking the link with the labour theory of value as a practical tool for the pricing of forest resources.

Stumpage tax. One of the problems with such a system of price formation is that it is very difficult to define the 'socially necessary labour input' in the production of naturally occuring forest. Elsewhere the author has pointed out that the pricing of naturally occurring timber at a level equal to the costs of exploitation will not lead to optimum decision-making regarding reafforestation unless a set of highly unrealistic assumptions are made (Blandon, 1978). If the price of timber in the Soviet Union were set equal to just the cost of felling and transporting, there would be an insufficient cash flow generated to pay for the costs of reafforestation.

Partly to overcome this, a stumpage tax was introduced which accounts for about 10 per cent of the costs of timber production. For the purposes of stumpage tax, the country is divided into seven belts and the tax is differentiated between belts. The tax within each belt depends upon a number of factors, the extraction distance, the species mixture, the average stem volume for example. The tax is higher in those belts which have few natural forests and lower in the north and east of the country. This is to encourage the concentration of production in the heavily forested areas. Within a given belt, however, the tax is higher if the extraction distance is short and if the trees are of a high average volume. This will have the effect of equalising costs across *lespromkhozi* operating in a given belt (Sudev *et al.*, 1976).

The total tax bill is calculated by surveying the areas to be logged and multiplying the standing volume by the relevant tax rate. Apart from acting as an incentive to concentrate logging in the heavily forested areas rather than the lightly forested ones, the tax has two other aims. One is to encourage the full use of the growing stock. Once the tax has been paid the cost is incurred regardless of whether the trees are felled or not, so it would be more economic, all other things being equal, to fell all the trees on one site rather than incur a higher tax payment by moving into another area. However, as we have seen, all other things are not equal and this aim of the tax has not been realised. Secondly, the tax was supposed to cover the costs of the sylvicultural activities required to regenerate the timber that was felled. This it has also failed to do with the costs of sylvicultural operations exceeding the tax receipts for many years now. Indeed, the stumpage tax of about 10 per cent of costs is one of the lowest in Eastern Europe. In the German Democratic Republic it is 55 per cent and in Hungary between 20 and 25 per cent (Lazarev, 1979).

Arguments are put forward in the technical journals for an increase in the tax along with a modification of the price structure in order to redress the situation.

> 'Of course, an increase in the tax ought to be accompanied by an increase in the prices of round wood and other timber products. It is necessary that the tax on timber as well as the prices on round wood reflect the socially necessary expenditure of labour ... The low level of taxes have not, for many years, covered the costs of the sylvicultural operations. As is well known, the deficiency of funds for those involved in sylvicultural work is covered by the State budget. The practice of making good part of the expenditures from the profits of other branches of the economy

> does not accord with the principles of socialist economics.'
>
> Lazarev, 1979

The author of this quotation implies that the 'socially necessary labour' required in the logging industry is that which is needed to replace the trees felled - their replacement cost and not just their cost of exploitation. At the present not only is the cost of timber below this (as illustrated by the fact that the stumpage tax is less than the expenditures on sylviculture) but also the sylvicultural operations actually carried out even with contributions from the State budget are insufficient, as was seen in Chapter 3. If the suggestions of Lazarev are taken up, the price of timber will rise sufficiently to cover the costs of regeneration and tending of plantations and this will be to the long-term benefit of Soviet forestry.

8
Individual Economic Problems

The discussion of price formation in Chapter 7 showed that prices in the Soviet Union are determined independently of considerations of relative scarcity and not allowed to fluctuate, being fixed for long periods of time. Thus, they are poorly suited to the role of acting as signals for the correct allocation of resources. For example, the choice between establishing either a spruce or pine plantation could not easily be made by reference to the prices at which spruce and pine logs are sold to the consumer enterprises, as would be the case in the West. There is no guarantee that the relative prices, fixed by the planning agencies, reflect the relative desirability of the two types of timber to the national economy and therefore no guarantee that the 'correct' species would be chosen.

In one important case, however, a price is totally lacking. There is no market for investment funds in the Soviet Union. The theory of market economies suggests that the rate of interest is determined when the market for funds for investment and savings is in equilibrium. At the equilibrium rate, savers are willing to forego some present consumption for increased future consumption, i.e. consumption equalling their principal and interest earned, and investors will only invest in those projects whose rates of return exceed this rate of interest. In this way consumers, by deciding how much to save at any given rate of interest, are able to express that 'rate of time preference' which they find acceptable, that is the increase in future consumption which is necessary to induce them to save money and so forego consumption in the present. Similarly, investors will not tie up funds by investing in projects yielding a return lower than this rate of time preference. In this way investors' decisions are consistent with the desires of the consumers.

This, at least, is what occurs in theory and it provides the rationale for discounted cash flow techniques which are the basis of the definition of the optimum rotation length and other investment appraisal techniques used in the West. Of course, there are problems in the application of this type of approach which are well recognised by Western economists but the absence of a market determined interest rate does pose special problems for forest economists working in the Soviet Union.

In this chapter we will look in detail at two of these problems and examine the methods used to reach decisions in order to illustrate the similarities and differences when compared to Western approaches.

THE EFFECTIVENESS OF NEW TECHNOLOGY

One problem now being tackled by Soviet forest economists lies in the evaluation of new technology and methods. As was described in Chapters 4 and 5, the methods used in logging are undergoing a radical transformation; there is, therefore, a clear need for a set of criteria to evaluate what constitutes the 'best' logging system under various sets of circumstances. This choice has to be made between a whole range of alternatives (feller-skidder, feller-buncher, and grapple skidder, chain saws and a choker and winch equipped skidder, etc.) and in Chapter 5 it was seen that, from the technical point of view, of all these systems, the feller-buncher and grapple skidder combination was emerging as the ideal system in a wide range of situations. However, from the economic point of view, the position is less clear. The peculiar situation has arisen whereby the logging system chosen as 'best' by the accepted economic criterion may, in fact, produce results counter to the intentions and objectives of the planning agencies. But in order to understand the criteria used in such choices at present, it is necessary to understand the development of capital appraisal techniques in Soviet economics.

Development of Investment Appraisal Techniques

After the revolution, one of the long-term problems facing the Bolshevik Government was that of inducing economic growth in the Soviet Union. Part of this desired growth was to be achieved by the use of new capital and new techniques, so there was a need to assess the value of new methods and investment in new equipment so that the scarce resources of the post-civil war Soviet economy could be directed to those uses where there would be the maximum effect. In the early 1920s there was quite a vigorous debate on how to measure the effectiveness of new capital that included discussions on methods involving calculations similar to discounted cash flow techniques.

However, Soviet economists were hampered in their attempts to evaluate investments by two factors. First, according to Marxist economic theory only current and past labour can create value. Thus, the only charge that could be made for the use of capital equipment was depreciation - representing the use of the past labour embodied in the equipment. This was clearly an amount fixed by the life of the good in question and so independent of the relative scarcity of capital goods. This had the effect of making capital appear cheap to its potential users and so encourage its extravagent use. Secondly, there was no recognised way in which the time factor involved in investment could be taken into account. As was just mentioned, there were until 1926, suggestions about a discounting type of calculation similar to that found in Western economic theory. But the very similarity of the interest rate calculation to the speculative activities of capitalist economies made such suggestions very unpopular.

These two factors meant that there was no real methodology for assessing those projects which maximised the return on capital and so no way in which capital could be allocated to those areas where it would be most productive. In Stalin's time debate in such matters ceased and some of the discussants were imprisoned. Capital began to be distributed administratively; this was partly a result of Stalin's desire to increase the control from the centre in economic life and not allow any 'objective' formulations to contradict his ideas for the development of the Soviet Union. It is likely that any form of calculation would have shown that investment in light industry would have yielded a faster, greater return than investment in other branches of the economy. But Stalin's ideas of economic development ran counter to this, demanding investment in heavy industry at the expense of light and consumer-oriented industries.

For logging, these developments had little direct relevance, but in other industries the choice of technique was more real. Here, unofficial allocation rules were springing up and these were finally made official in 1960. In that year *Gosplan* and the Academy of Sciences jointly published a booklet describing the methods recommended for use in choosing between alternative technical methods in investment (Anon, 1960). The technique suggested was a formalisation of unofficial methods that had been in use in the project testing institutions in the railway and electrification industries. The method involved the use of a *coefficient of relative effectiveness* (CRE). The CRE was used to decide between two alternative strategies and is given by the equation below.

$$\text{CRE} = \frac{C_2 - C_1}{K_1 - K_2}$$

Here C_1 and C_2 are the annual operating costs of systems 1 and 2 respectively and K_1 and K_2 the respective capital outlays. We would expect that, if C_2 were greater than C_1, K_1 would be greater than K_2. If, for example, system 1 had lower operating costs and also lower capital outlays, there would be no need for a decision-making criterion. Referring to the equation, the numerator shows the annual saving in operating costs that would be made if system 1 were used instead of system 2, whilst the denominator shows the extra capital cost incurred in the purchase of the capital equipment.

The reciprocal of the CRE would show the number of years required for the extra capital expenditure needed for system 1 to be recouped from the saving in the annual operating costs - ignoring any discounting aspects of the timing of the cash flow. This is similar to the simple payback period criterion - the length of time before the amount of an investment is recouped from the increased profit which it generates. Actually, its relationship to the simple payback period is of interest as it shows some important features of the investment decision in the Soviet Union, but this will be discussed later.

From inspection of the equation it should be clear that the greater the CRE the better will be system 1 compared to system 2, for the greater will be the savings in operating costs for a given extra capital outlay. To make use of this criterion, a norm was established and the decision to choose system 1 would be made if the calculated CRE exceeded this norm. For the forest industries the norm was established at a value of 0.12. This suggests that the payback period, or the number of years that system 1 has in order to recoup the extra capital expenditure associated with it is 8.3 years

- the reciprocal of 0.12.

The norm varied according to the branch of the economy being considered and its priority in the overall plan for the development of the economy. The greater the norm the more effective would investment have to be in that particular branch of the economy. A given increase in capital outlay would have to return a greater saving in costs. A lower value for the norm would both permit and encourage increased investment in that part of the economy. In practice, the range of values for the norm was from 0.33 in light industry to 0.05 in the electricity generating industry thus permitting an increase in investment in electricity generating capacity at the expense of light industry. Clearly, moving funds from investment in branches of the economy with low norms to those with high norms would yield a net gain to the Soviet economy in the short-term at least. But this movement of funds was not what the Soviet planners had in mind when they viewed the growth of the economy and so these differentiated norms were viewed by the planners as an indirect means of administratively allocating capital.

The value of the norm used in the forest industries was unremarkable in that it represented the approximate all-industry average norm for the time, showing that the planners did not view logging and wood processing as a branch of the economy to be explicitly encouraged or discouraged.

Some authors conclude that the average norm was set too low to ration capital. That is, all capital when used in investment did not depress the return on capital below that value implied by the norm. So, they conclude, administrative rationing of capital remained the usual practice (Thornton, 1971 and Gregory and Stuart, 1974).

In 1969 the approved methodology was slightly altered to make its use easier when choosing between more than two ways of achieving the same end. If En is the norm of effectiveness of capital, then system 1 will be chosen if the following equation is true.

$$En < \frac{C_2 - C_1}{K_1 - K_2}$$

or

$$C_1 + EnK_1 < C_2 + EnK_2 \qquad (1)$$

From this formulation of the criterion the rule was restated as requiring the choice of the system for which the value of C + EnK, called the *calculated cost indicator* (CCI), was a minimum. With this method the calculations are usually carried out on a unit output basis so that the total annual operating costs and the capital costs are divided by the annual output of the system in question (Anon, 1969).

Many Soviet and foreign commentators had noted that the differential norms led to the possibility that overall output could be increased by a reallocation of capital away from industries with a low norm to those with a high one - as was noted earlier. Estimated losses associated with these differential rates varied from as little as 3 or 4 per cent per annum to over 30 per cent (Thornton, 1971). So, at least initially, with the introduction of this modification to the capital appraisal methodology a uniform norm was suggested for all branches of the economy of 0.12. Variations in this figure were allowed only for investment in remote areas and for very long-term projects with the agreement of *Gosplan*. This uniformity in the norm throughout the economy showed the move on the part of the authorities away from an administrative scheme of capital allocation and a move towards the ideas of the 'optimal planners' - for a fuller description of this see Ellman (1971 and 1973).

However, by 1973 differential norms had been re-introduced so strengthening the role of the planning organs and reducing the effectivensss of these new, 'objective' economic criteria. Those norms applying to the logging and woodworking industries are given in Table 8.1. The average figures for these industries as a whole was 0.12 and 0.13 respectively - again near to the all-industry average.

TABLE 8.1 Economic Effectiveness Norms Introduced in 1973

Industry and Branch of Industry	Norm of Capital Effectiveness, En
The Woodworking Industry	0.13
sawmilling, plywood manufacture	0.11
match manufacture	0.13
joinery products	0.13
fibreboard, particle board manufacture	0.09
prefabricated housing units	0.13
furniture manufacture	0.15
The Logging Industry	0.12
logging	0.11
road and rail transport	0.11
timber transport by water	0.13

Source: Mitin and Sarkisov, 1974

In the woodworking industry it is interesting to point out that the norm varied from 0.09 in particle board and fibreboard manufacture to 0.15 for the manufacture of furniture. This differential, as was described earlier, would be designed to encourage investment in particle board and fibreboard plant at the expense of investment in the manufacture of furniture. Reasons for this are probably two-fold. First, from the point of view of export earnings, particle board and fibreboard are more important to the Soviet Union than furniture and so investment in the former was to be encouraged so as to provide Soviet industry with a wider base for its forest products' exports. Secondly, with the emphasis on the utilisation of as much of the tree as possible and the resulting increase in chip production, an increase in output of those products that use chips was required. In the logging industry logging was given a norm of 0.11 whilst log floating and water transport of logs a norm of 0.13. This suggests a slight emphasis on logging and investment in transporting techniques other than floating.

In this new official methodology a greater emphasis was put on the techniques of discounting in evaluating projects in which the capital payment continued for more than one year. The discounting factor is of the normal type expressed in the equation below.

$$\text{Discounting factor} = \frac{1}{(1 + Ed)^{t}}$$

Here, t is the time when the capital expenditure occurs and Ed, the discounting norm was set at 0.08, or 8 per cent per annum. Although only implicit in the original publication (Anon, 1969) this type of calculation was intended to be used for those investments with expenditures occurring over only very short time spans, i.e. when t is fairly small. The use of discounting techniques over very long time periods causes many problems for Soviet forest economists, as will be seen in the next section of this chapter when the question of the determination of the rotation will be considered.

But this more explicit acceptance of the time factor in the investment decision should have led to an appreciation of the variation of output from a system over time. For example, in Chapter 5 it was shown that the introduction of the new logging machines leads to an initial fall in output of the logging brigades below the level of the planned output. This effect lasts until the crews become accustomed to the new methods and only after this time lag does output rise to the planned levels - although often it fails to do so. Clearly, the loss of output in the early stages of a new system represents a cost and should be included in the investment calculations. In Mitin and Sarkisov's book (1974) a loss of output during the rebuilding of a factory is restated in terms of a loss of profit and this is added to the capital costs of the relevant year to be discounted. However, in their book this method is applied only to such problems as the rebuilding and re-equipping of lower landings and sawmills and not to the introduction of new logging systems. Recently some allowance seems to have been made for this effect in the introduction of radically new systems by the advocating of a discount rate of 10 per cent for investment in systems already in use elsewhere, whilst 8 per cent is to be used to discount expenditures in capital of a new, qualitatively distinct type (Yanushenko, 1977).

It could be argued that there is a link between ignoring the effect of time on the level of output of a particular system and the ignoring of the idea of risk in Soviet planning. This point has been touched upon in the last chapter in the discussion on the nature of the planning system. If there is a more explicit acceptance of the idea of time in the evaluation of different investment alternatives, instead of using a criterion that assumes that a system provides an average annual output throughout its life, then risk intrudes into the analysis. In the West, a more 'advanced' logging system may be chosen by an entrepreneur, not because of its immediate return, but rather because of his 'feel' that the machines will provide a higher return in the future. Alternatively, a producer may decide to manufacture the new types of skidder because of his 'feel' that demand will take off and he will then be in a position to capitalise on an expanding market.

Such ideas are formalised in Western managerial economics into methods of risk analysis and techniques for decision-making under uncertainty. In the Soviet Union a similar source of risk and uncertainty clearly exists. A new machine or system is likely to be subject to unforeseen problems or potentialities regardless of the economic system, and so future economies to be gained by its use are to a certain extent unknown. However, in the Soviet Union with its emphasis and pride in the planning of the economy, there is little emphasis on the risk and uncertainty attached to investment decisions. Indeed, future risk implies the acceptance of some form of planning failure, so the absence of risk analysis in such an explicit form as it appears in the West is not surprising.

Problems with the Appraisal Techniques

The reciprocal of the CRE introduced in 1960 measures the payback period of the *extra* capital outlay of the more capital intensive system in terms of the reduction in operating costs that it produces. In the West, of course, the payback period of a system measures the length of time before the *total* capital outlay is recouped. The apparently peculiar Soviet formulation makes it possible for a system to be chosen which fulfils the CRE criterion but results in the introduction of a system which has a payback period, in the Western sense, in excess of the norm value.

Imagine an initial logging system, using chain saws and an ordinary skidder, with annual operating costs of 16,000 roubles. Let two alternative ways exist of modifying the skidder, costing 2,000 and 3,000 roubles for the required parts and fitting, and the new operating costs, with the improved tractor, be 16,000 and 15,800 roubles respectively. These cost savings of 100 and 300 roubles per annum would give normal payback periods of 20 and 10 years - outside the norm of 8.3 years. However, the CRE comparing improvement 1 with improvement 2 would give a value of 0.2 or a payback period of 5 years - within the value of the norm. Thus improvement 2 would be chosen and implemented in preference to improvement 1 as the greater capital expenditure would have been justified.

The example shows that the CRE, and the *calculated cost indicator* which derives from it, were designed not to choose which type of investment was economically most effective, but how best to follow a particular policy of investment once the decision to invest in a particular way had been taken by the central planning authorities. Absolute norms and methodologies to go with them do exist to establish whether an investment programme is worthwhile on its own merits. However, these tend to be used only on the macro-level and not on an individual basis. For example, in the case of logging, the decision to increase the productivity of labour has been emphasised many times by the planning authorities for reasons discussed in Chapter 6. So the question that these criteria have had to answer is how best to increase the productivity of labour and not whether it is economically 'correct' or 'incorrect' to set about increasing labour productivity by an influx of capital. Here it is useful to quote Varaksin and Stupnev (1974) at length.

> 'Here it is opportune to turn our attention to the differences in the approach to the problem of mechanisation (in

the East and the West). If, in our country and other socialist countries questions of mechanisation are an integral part of the general programme of improvement of working conditions and the increase in the well-being of the people, then in the countries of the so-called West the process is brought about and defined by purely economic considerations.

As a result, relatively complex and thus expensive multi-operation machines for the complete mechanisation of logging ... even under two or three shift work regimes are unable to compete with traditional methods of logging. These conclusions apply only to countries in the West with high (relative) prices for capital equipment.

This is why, in many countries where wages of labour are low, ... there is no attempt to mechanise the logging process and it is often possible to see felling and even cross cutting being carried out by axe. Indeed, several Western economists link the optimum level of mechanisation with the possibility of the use of manual labour. The forest industries of our country are developing along different lines. The mechanisation of logging is an integral part of the growth of industry.'

So the labour-intensive methods have been eliminated from the list of possible investment strategies. This contrasts with the Western economists' view that the optimum capital intensity would be dependent upon the relative prices and productivities of capital and labour and so labour-intensive methods are 'correct' in areas with a capital scarcity and an abundance of labour. However, a cynic might point to the labour shortage as being the stimulus for logging mechanisation as it would be in the West.

But the *calculated cost indicator* is failing on two counts to provide the correct guidance in how to mechanise logging, first from the point of view of tractor builders, where the criterion is in fact having a perverse effect, and secondly, from the point of view of the logging industry itself.

Tractor production. From the point of view of the managers of the factories producing tractors for the forest industries, there seems to be little incentive to produce new forms of logging equipment. In investing in new lines of production, investment analysis for the manufacturers of tractors is carried out in the same way as was described earlier. The factors taken into account are those which enter into the *calculated cost indicator* for the tractor producer, in this case the operating costs and capital costs of a new production line. As Rodigin (1977e) writes;

'The machine builders are only worried about lowering their *calculated cost indicator*, and they are not answerable for the productivity of labour in logging.'

So a factory producing tractors, having received in the plan instructions to produce new forms of tractor, will find itself encouraged by the indicator to produce that form of new tractor which has the minimum operating and capital costs in its production run. It is likely that this condition will be fulfilled by the modification and upgrading of an old type of tractor already in production rather than the installation of a line to produce a radically new type of machine. Therefore, 'it is not surprising that old technology is often produced under a new name' (Rodigin, 1977e).

In recognition of this problem - which occurs in other branches of the economy - a new indicator of economic effectiveness was introduced in 1977 (Anon 1977d). Recalling equation 1, the annual economic effect (AEE) of using system 1 rather than system 2 may be defined by the equation below.

$$AEE = \big((C_2 + EnK_2) - (C_1 + EnK_1)\big)Q_1$$

Here the *calculated cost indicators* are worked out on a unit output basis and Q_1 is the output of the new system 1. This may be rewritten as the equation below where CCI represents the *calculated cost indicator*.

$$AEE = (CCI_2 - CCI_1)Q_1$$

In the case of tractor production, systems 1 and 2 may be interpreted as two opportunities of producing new types of tractors. System 1 is the more advanced system involving the production of, say, a feller-buncher or a feller-skidder, whilst system 2 is the less advanced alternative of producing a slightly modified variant of a type of tractor already in existence, for example, a winch and choker equipped skidder of slightly increased power. CCI_1 and CCI_2 in this case would be the *calculated cost indicators* associated with systems 1 and 2 and Q_1 the number of tractors produced under alternative 1 by the tractor factory.

As was argued earlier, it is likely that CCI_1 will be greater than CCI_2 and so investment 2, the modified tractor variant, would be chosen and the annual economic effect, as defined by the equation above, would be negative - representing a saving.

However, the new indicator redefined the annual economic effect so that the increased productivity and service life of the newer tractors would be taken into account as well as reduced operating costs to the tractor users. According to the new instructions, tractor producing factories (and, of course, producers of other capital equipment) should now evaluate the annual economic effectiveness of the production of new equipment by equation 2.

$$AEE = (CCI_2.a.b + U - CCI_1)Q_1 \qquad (2)$$

In this equation, which is presented by Zaitsev (1978) without the benefit of derivation, a would be the growth of productivity of the new tractor compared with the modified old type when in use in forests, b the proportional change in the service life of the new type of tractor compared with the modified old tractor and U would be the users' savings in operating costs whilst using the new tractor instead of the modified variant calculated on a per tractor basis.

Thus, if the new tractor to be produced under investment alternative 1 has twice the life and twice the productivity when in use of the modified tractor produced under alternative 2, both a and b would be equal to 2. This would increase the economic effectiveness of producing the more advanced and more desirable form of tractor. So, at first sight, this criterion would seem to be what is required to induce the producers of tractors to produce the desired form of new machine.

But a closer examination of the situation reveals that this is not so. Factor a refers to the machines' productivities and not to labour productivity. Whilst labour productivity in logging might rise quite a lot on the introduction of the new technology, machine productivity does not. For example, the LP-2, LP-17 or VTM-4 often do not harvest a significantly greater volume of timber per machine shift than does a TDT-55 or 75 in conjunction with a chain saw, so factor a will not differ greatly from unity. The factor b will also be close to 1 in its value, indeed, if anything it may be less than 1 as the new machines (being more complex) are likely to be less reliable and so have a shorter life span. The savings for the tractor users, U, is also likely to be negative because of the higher amortisation costs and wage payments to operators associated with the new technology.

So there is every chance that, overall, this new formulation of the *calculated cost indicator* will act even more perversely than the original indicating, for the producers of tractors, the production of only modified variants of the old types of machine and not the mass introduction of the more desired types of radically new machines to mechanise completely the logging process.

Another factor related to the planning system discussed in the last chapter conspires to produce the same effect. Stupnev, First Deputy at *Minlesbumprom* wrote in *Pravda* (quoted in Rodigin, 1977e)

> 'Having worked out solutions to the technical problems, which allow us to increase the productivity of labour several times, we are obliged to use obsolete equipment. [The reason is that] the machine builders are not interested in the introduction of new machines. The old models, they say, are profitable to produce ... '

Profitability, as was noted in Chapter 7, is an important indicator and the introduction of the production of a radically new design of product is nearly always associated with a decline in profitability, at least in the early stages of production. In general, the prices of new capital equipment are set too low to increase the profits of the producers. Figures for tractor production illustrate this point. The rate of profit as a percentage of cost on the production of the older types of tractors, such as the T-100, was between 20 and 30 per cent, whilst the figures for the newer agricultural tractors, such as the T-4 and T-40, were between -70 and -3 per cent. However, as Zaitsev (1978) points out, in the Soviet economic system of price formation 'there is no economic justification for establishing the price of a new product (to give) a higher profitability level than on the old product'.

This point is also indicative of high production costs of the new type of logging equipment. Absolute profit is defined as the difference between price and production costs multiplied by the level of output. The price of the new types of logging machines such as the LP-2, LP-17 or VTM-4 is around twice or one and a half times the price of the older machines. Yet, if these older machines are as profitable or more profitable than the new ones, the implication is that the production costs of these new machines are correspondingly greater. The differential diminishes through time, but not alarmingly so. The Krasnoyarsk tractor factory has reduced costs on new forestry tractors by only 30 per cent over a six-year period (Shtrek, 1978). This reinforces the view put forward earlier that the *calculated cost indicator* for producing machines that are only modifications of older types is considerably less than the indicator associated with the production of the newer types of logging equipment.

The two effects have led to the following paradoxical position. Since 1947, during which time an increase in the productivity of labour has been continually called for so as to improve working conditions, the tractors introduced have not altered radically in the way in which they operate. Only their power has increased, after the diesel engine had come to the fore. The procession of tractors - KT-12, TDT-40, TDT-40M, TDT-55, TDT-55M, TDT-60, TDT-75, TT-4 - have all been equipped with winches and chokers. The logging process has remained essentially unaltered but the systems' productivities have increased from about 40 to about 70 cubic metres per shift. This is an increase in the productivity of labour on the forest floor of about 1.7 per cent per annum which compares unfavourably with the rest of the forest industry and the rest of the economy. Yet the workers on the forest floor now work harder than they did before, especially the chokerman whose job has remained unaltered during the entire period. He (or she as several thousand women work as chokermen) must now connect chokers to 70 cubic metres per shift rather than 40. This is surely a decrease 'in the well-being of the people' recalling the words of Varaksin and Stupnev quoted above.

But when the new machines are introduced it is not all benefit to the forest workers. As has been mentioned before, the high amortisation payments associated with the high prices of the new machines have led to the introduction and extension of multi-shift working regimes in order to spread these fixed costs over a greater volume of timber. Yet this means that work is done during the less convenient parts of the day - worsening the working conditions. But the tractor producers would probably complain that the price of new tractors is still too low for them to maintain their level of profitability if they opt to produce new rather than old types of machine. But the prices now ruling are not only producing the effects just mentioned but are also causing investment appraisal techniques in the logging industry to fail to indicate the most desirable forms of logging mechanisation. This is causing some forest economists to attempt to redefine the nature of economic effectiveness.

Appraisal of new logging methods. In the appraisal of new logging methods in forestry the *calculated cost indicator* is also failing to provide guidance. The capital costs of the more advanced systems are higher. Also, the operating costs tend to be higher, for the new machinery has higher amortisation and servicing costs and, although fewer workers are employed, those who are required are usually better qualified and, therefore, paid more. But the *calculated cost indicator* is expressed in terms of the output of the system (roubles per cubic metre) and so the productivity of the system is the important factor and not the productivity of labour.

Rodigin (1977e) shows the anomalies that this causes. Looking at skidding, he compares a basic skidder with a winch and choker log gathering system with two possible alternatives - an improved winch and choker equipped skidder and a grapple skidder. Table 8.2 shows an example very similar to Rodigin's. The TT-4 is cheaper to purchase than the TB-1 because it is a simpler machine, an adaptation of an agricultural tractor equipped with a winch. The TB-1, although based on an existing tractor, is a grapple skidder and so has a more complex log gathering system which is reflected not only in its initial cost but also in its maintenance costs.

TABLE 8.2 Economic Effectiveness of New Skidders

	TDT-55	TT-4	TB-1
Cost, roubles	6,000	8,000	10,000
Number of workers required	2	2	1
Output per shift, cubic metres	60	90	70
Wages per shift, roubles	24.40	24.40	14.04
Maintenance and running costs per shift, roubles	27.53	36.00	38.62
Operating costs per shift	51.93	60.40	52.66
Operating costs per cubic metre	0.87	0.67	0.75
Capital costs per cubic metre (En = 0.12)	0.04	0.04	0.06
CALCULATED COST INDICATOR	0.91	0.71	0.81
Change in the calculated cost indicator	-	0.78	0.89
Change in labour productivity	-	1.50	2.33
EFFECTIVENESS INDEX Er	-	1.83	2.59

As a result, the *calculated cost indicator* would indicate the TT-4 is more desirable than the TB-1 so, at the same time, prolonging the need for a chokerman and increasing the work that he has to do. To overcome this, Rodigin attempts to reformulate the measure for economic effectiveness. To achieve the desired results, he redefines the various parameters in terms of labour rather than output units. The exposition of this criterion is not clear but it does illustrate the problems that are facing economists in this field in the Soviet Union.

Let P' be the change in the productivity of labour between the existing system and a new system, and E' be the change in the *calculated cost indicator* between the existing and a new system. These figures have been calculated in Table 8.2. Rodigin defines the benefit that accrues to each member of the work-force from the expenditure on the new equipment as P' - P'E'. To take into account the reduction in the number of workers required because of the increase in labour productivity, Rodigin adds the increase in the productivity of labour, P', to the already defined economy accruing to each member of the work-force to give a figure for overall effectiveness, Er, defined as

$$Er = P'(2 - E')$$

There seems to be little theoretical justification for this particular formulation of the criterion and the mixture of units represented in the formula makes the equation difficult to interpret. However, the criterion is evaluated for the two choices of skidder in Table 8.2. From this the TB-1 emerges as superior to the TT-4, a result more in line with the desires of the planning agencies.

The very *ad hoc* nature of this criterion indicates the problems faced in this field of Soviet economics. The element (2 - E') in the definition could become negative and so produce meaningless results. On this point Rodigin can only say:

> 'If the *calculated cost indicator* increases by two or more times, then even the most highly productive machines will appear ineffective. In any case, the increase in the *calculated cost indicator* should be minimal whilst the growth in the productivity of labour should be as great as possible.'

In another suggested approach to the problem, Burdin (1975) proposes a two-tier price system for capital equipment. The economies made with the use of the new equipment are shared between the user and the producer. The producer receives a high price to cover the production costs and the users are required to pay only a low price - the difference being met, presumably, by the State. This allows the depreciation costs to be correspondingly reduced. Also, he suggests that a fund be set up for the use of the users of the new machines during their introduction period to help cover such costs as the training of the operators, the necessity to buy large stocks of spare parts and other extra expenditures incurred in the initial stages of using the new technology.

These two approaches show both the theoretical and practical problems being created by the Soviet economic system in the mechanisation of logging. Theoretically, the guidance being offered by economics as to the choice and development of the various systems is suspect and in practice the price system and investment criteria are hindering the large-scale introduction of the new machines.

The criterion to evaluate the effectiveness of new technology is used in almost every area of investment appraisal, from the effectiveness of methods of nursery work to the determination of the best size of logging enterprise. However, one area where it is not used is in the definition of the optimum rotation length in Soviet forestry. Here problems of a slightly different nature are being encountered.

THE DEFINITION OF THE ROTATION LENGTH

The definition of the rotation length is, after the choice of species, arguably the most important decision to be made in forestry. In the West, the rotation length is usually arrived at by using discounted cash flow methods - originally suggested for use in forestry in Martin Faustmann's work of as long ago as 1849. Indeed, in the

West the main controversy has been over the correct way to use the idea of discounted cash flow techniques - whether it is the net present value or the internal rate of return of the cash flow that is to be maximised.

In the Soviet Union each year the timber from about 3 million hectares is harvested. With a total forest cover of about 1,200 million hectares this would imply a rotation length of about 400 years. So there would seem little danger of over-cutting and so no need to expend too much research effort to define a precise rotation length from carefully defined premises.

This over-simplification, of course, ignores the problem of location. As was seen in Chapters 2 and 3, much of the 1,200 million hectares of forests are remote, reserve forests, not as yet being fully exploited, whilst the European forests are being over-cut. The over-cutting in Europe is being matched by under-cutting in Siberia. Gradually, the weight of logging operations is being relocated and this means that Soviet loggers are having to come to grips with the problems that this poses. In response to this, there has been a growing interest in the correct definition of rotation lengths since the 1960s - partly out of purely academic interest and partly so as to manage better the more accessible timber growing areas and so reduce the severity of the unfortunately inevitable problem of transport of timber from remote areas.

The Present Situation

At present 'The Methodology of the Calculation of the Allowable Cut in the State Forests of the Soviet Union' is used in the formulation of a definition of the allowable cut - a figure which the logging concerns must strive to achieve. In some areas, these instructions define the rotation length quite precisely, for example in the Leningrad region the rotation length for spruce is 81 years whilst for birch it is 61. In other areas of the country a large range of possible values is allowable which reflects the fact that in the over-exploited areas of Russia the rotation length and therefore the allowable cut have to be subject to a greater precision than in the under-exploited regions.

The 'Methodology ...' advocates the use of technical rotations in forests of Groups II and III. The concept of a technical rotation is shown in Fig. 8.1 In this diagram, curve I shows the growth of volume of timber on 1 hectare of land. Typically this will be a sigmoid curve with a maximum mean annual increment at age T_1 At that age, a ray from the origin of the graph (the slope of which represents the average annual growth rate) touches the growth curve so as to give the maximum slope. This age, T_1, represents the maximum average annual growth rate of ALL timber on a site. However, not all of this volume of timber will be useful; some of it will be too small. So the technical rotation is defined as the rotation length which maximises the average annual increment or growth rate of the most useful or desirable timber sizes. This is also shown in Fig. 8.1 To produce any of the desired timber sizes at all, t_1 years are required, and after that, the increase in volume of the desired sizes of timber follows a similar sigmoid curve to that describing the entire volume on the site. This is curve II in Fig 8.1.

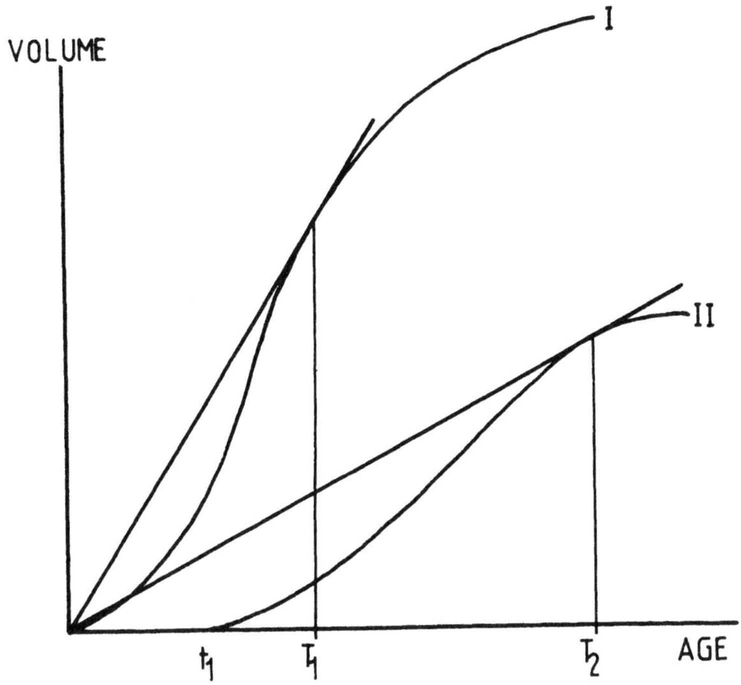

FIG. 8.1 The Technical Rotation

The maximum annual increment of the desired timber sizes (the technical rotation) is T_2.

The main objections that Western economists would make to the use of the technical rotation in forestry is that the timing of timber production, and therefore the timing of the financial receipts and expenditures, is ignored. This objection, it would be argued, is overcome with the use of discounted cash flow techniques. The technical rotation is an almost purely physical or biological criterion, except for the fact that the desired timber assortment used in its definition would, presumably, refer to some economically desired timber assortment.

It is not really surprising, however, that this should be the case in the Soviet Union. Until 1966 the plans that forestry concerns had to comply with were largely defined in physical terms. Even now, after the reforms of 1966 that were supposed to allow price and value categories to play a more important role in the formation of the plans and the evaluation of performance, the dominant plan indicators remain the ones dealing with volume. In this environment it is not surprising that the definition of the rotation length remains one that is essentially physical or biological in nature rather than one which deals explicitly with economic concepts.

The Applicability of Discounting Methods

In most Soviet forests, as has been explained, the basic operation is one of logging the stands of timber coupled with some measures to re-establish the crop. In this situation the argument about whether the rotation is an economic or biological concept may be irrelevant.

If the volume of desired timber assortments produced from a given area is a function of time such as that described by curve II in Fig. 8.1, it may be written as $V = V(t)$, where V is the volume per unit area and t is the age of the trees. If the aim of maximum sustained yield is accepted, the annual yield from the forests of the Soviet Union will be v given in equation 3.

$$v = \frac{C.V(t)}{t} \qquad (3)$$

In this equation, C is the total accessible forest area in the Soviet Union, which will vary through time. If, however, it is assumed that C is constant, at least in the short run, the maximisation of v becomes simply the problem of maximising $V(t)/t$ with respect to t. Differentiating equation 3 with respect to time gives equation 4.

$$\frac{dv}{dt} = \frac{t.C.V'(t) - C.V(t)}{t^2} \qquad (4)$$

This should be set equal to zero and, assuming second order conditions for maximisation are fulfilled, this gives rise to equation 5.

$$V'(t) = \frac{V(t)}{t} \qquad (5)$$

This is the familiar condition that the rotation of maximum mean annual volume increment is that rotation length defined as time t when the mean annual increment, $V(t)/t$, equals the current annual increment, $V'(t)$. In terms of Fig. 8.1, the rotation length that maximised the output of some desired timber assortment is T_2. At T_2 the slope of the ray from the origin is maximised and is also tangential, i.e. has the same slope as, the growth function $V(t)$.

However, to return to the main point, if the objection is raised that the correct rotation length is an economic and not a technical question and that the true maximand should be a discounted cash flow stream and not the physical volume stream of timber, the function to be maximised is that shown in equation 6.

$$R = p.C.\frac{V(t)}{t} + p.C.\frac{V(t)}{t}\cdot\frac{1}{(1+r)} + p.C.\frac{V(t)}{t}\cdot\frac{1}{(1+r)^2} + \dots \qquad (6)$$

Here R represents the revenue to be maximised, p is the average price, net of costs, of the desired timber assortment, and r is the relevant discount rate. In perpetuity this forms a simple infinite geometric progression, which may be rewritten as equation 7.

$$R = p.C.\frac{V(t)}{t}\cdot\frac{1+r}{r} \qquad (7)$$

Here, the elements p, C and r may be taken to represent constants. In the Soviet Union this will not represent a too unrealistic assumption as costs and prices are infrequently adjusted. Thus, by varying the rotation length, t, so as to maximise R will only affect the factor V(t)/t in equation 7, which is effectively the same as equation 3. Therefore, the rotation length which maximises the net discounted revenue will also be that which maximises the annual volume production of the desired timber assortment, i.e. the technical rotation.

It might be argued that, as price is dependent on average tree volume the term for price in these equations should be re-written as a function of time. This would mean that the derivation of the rotation length that maximises net discounted revenue would not be so simple and so not necessarily equal to the technical rotation. The price of the timber assortments used to define the technical rotation may be sufficient to compensate for the effect of waiting for this timber to reach maturity. If the price of the desired timber sizes is not sufficiently great compared to the prices of smaller assortments, the rotation length that maximises net discounted revenue would be less than that which maximises volume production of the desired assortment. Also, if the price of the assortments produced if the trees were left to grow for longer than the technical rotation were of sufficiently greater value, then the rotation that maximises net discounted revenue would be the greater. However, in as far as the technical rotation is defined in terms of maximising the physical output of the most desirable timber volumes which should have a sufficient price to reflect its desirability, the argument above remains valid.

As well as this, the formulation assumes that a continual, uninterrupted flow of forest products is considered to be desirable. This is in line with current Soviet philosophy, although clearly the unconstrained maximisation of net discounted revenue or volume production without ensuring that output is continual may lead to other conclusions. In fact, equation 6 suggests that annual output is constant. Most Soviet experts would argue that the volume produced in Soviet forests should increase annually, and so the formulation in equation 6 is wrong. However, the allowing of output to grow or vary slightly should not affect the overall conclusions.

Finally, and possibly this is the most serious objection to the argument so far, equations 6 and 7 assume that the next generation of forests would be equally productive as the newly-felled one. This implies that control of the species composition of the new forests, or any site amelioration carried out, would have no effect. If future generations were assumed to be more valuable or more productive this would tend to shorten the rotation length calculated compared to those evaluated under the assumption of no change in productivity.

The Overall Significance of a Rotation Length

If natural forests are considered in isolation, the question of discounting cash flows and the problem of whether the correct rotation length is a biological or economic concept is, in certain circumstances, irrelevant. However, the use of technical rotations is still unsatisfactory from an overall viewpoint. As, in the Soviet Union, not only the forest but also the timber processing industries, transport and the industries which are timber users are owned by the

State, the forests should not be considered in isolation, as they would be, for example, by a Western forest owner with no interests in other branches of industry. Instead, the correct criterion to apply to the management of the Soviet Union's forests should be one which, in conjunction with correct criteria elsewhere, would lead to decisions that would contribute to the overall optimisation of the activities in the economy.

This is the approach that was suggested by the proponents of the 'optimally functioning socialist economy' referred to in Chapter 7. In his book 'Optimum Planning in the Forest Industries', Glotov (1973) notes the possibility of using two criteria in this context - a programming system involving the overall maximisation of profits, and one involving the overall minimisation of the *calculated cost indicator*, referred to in the previous half of this chapter. As a result of such calculations, the scale of logging would be arrived at by the maximisation of total profits in the economy, and decisions at the *lespromkhoz* level would be left to the *calculated cost indicator*. However, Glotov notes that the price system in the Soviet Union is not of sufficient sophistication to allow the use of such calculations. Indeed, the use of any 'economic' criterion has to face the problem of imperfections in the mechanism of Soviet price formation. Also, the mere complexity and size of such calculations have meant that, of late, emphasis seems to have moved away from such mathematical, programming techniques in the planning process.

The *LenNIILKh* Recommendations

One group do recommend the use of a criterion that goes some way to taking into account other branches of the forest industries, although being less ambitious than the global programming techniques suggested by others. Specialists at *LenNIILKh* (the Leningrad Scientific Research Institute for Forestry) have suggested that, not only should the correct rotation length be considered an economic concept, but also the cost structures of the wood-using industries as well as the logging industry should be taken into account in determining this rotation length (Krestiashin, 1974). Although their calculations are based on data for the Leningrad area and their recommendations concern the North-west region, the idea behind the method has general significance.

Their methodology is based around the relationships shown in Fig. 8.2. This diagram refers to spruce forests in the Leningrad region. Each of the variables described in the graph are shown as percentages of the maximum or minimum that they attain over time, the mean annual increment peaks, i.e. reaches 100 per cent of its maximum value, between the ages of 81 and 110 years. The usually recommended rotation length is the lower of this range, i.e. 81 years. Logging costs, expressed on a per cubic metre basis, decline to a minimum level (100 per cent of the minimum possible) when the forest is more than 150 years old. If the forest were logged over at the recommended rotation age of 81 years the diagram shows that logging costs would be about 15 per cent higher than the minimum level. Also, the sawing costs would be about 5 per cent greater than the minimum attainable.

To take into account the various aspects of the timber conversion

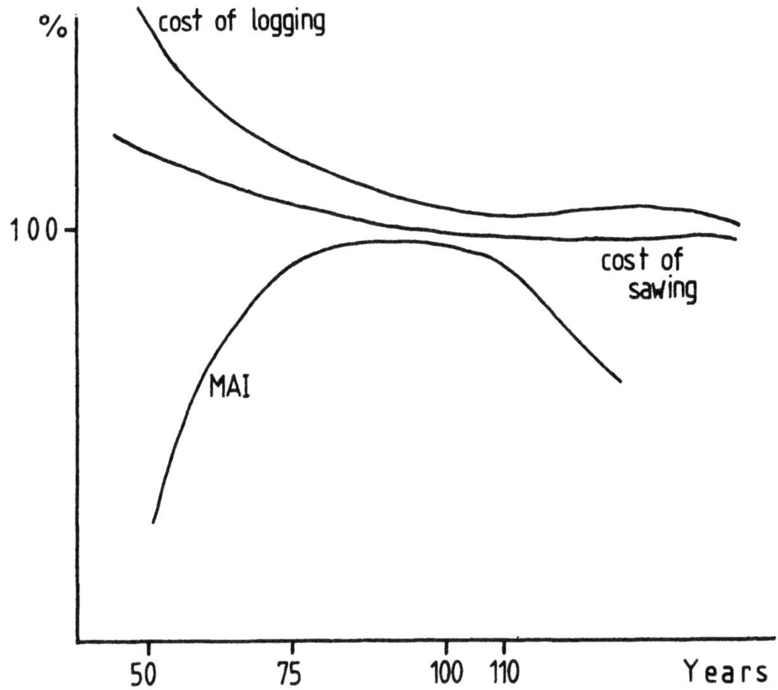

FIG. 8.2 The *LenNILKh* Recommendations

process, the author recommends that the optimum rotation length will be that which maximises P_t, the overall profit per cubic metre of timber defined by equations 8 and 9 below.

$$P = \left((R_1 - C_1) + (R_s - C_s)\right).I \tag{8}$$

$$P_t = P - \frac{(Q_1 - Q_2).Y}{A} C_i \tag{9}$$

Here,

P = the net average annual profit from sawmilling and logging the timber from 1 hectare of forest in roubles,

R_1 = the revenue obtained per cubic metre by the logging enterprises in roubles,

C_1 = the cost per cubic metre of logging in roubles,

R_s = the revenue obtained from the sawn timber products derived from 1 cubic metre of raw material in roubles,

C_s = the cost of purchasing and converting 1 cubic metre of timber into sawn products in roubles (note that the purchase price of the timber will not be equal to the sale price at the *lespromkhoz* - see Chapter 7),

I = the mean annual increment of the desired timber sizes per hectare in cubic metres,

Q_1 = the volume of timber in the total area that would have been felled if the technical rotation, i.e. the rotation length of maximum mean annual increment, had been used, in thousands of cubic metres,

Q_2 = the volume of timber in the total area that would be felled at the various ages under consideration to determine the optimum rotation under the *LenNIILKh* method, in thousands of cubic metres,

A = the total area under consideration in thousands of hectares,

Y = the yield of desired timber assortments per cubic metre of growing stock, expressed as a fraction, and

C_i = the cost of importing 1 cubic metre of timber from an exporting region into the area under consideration in roubles.

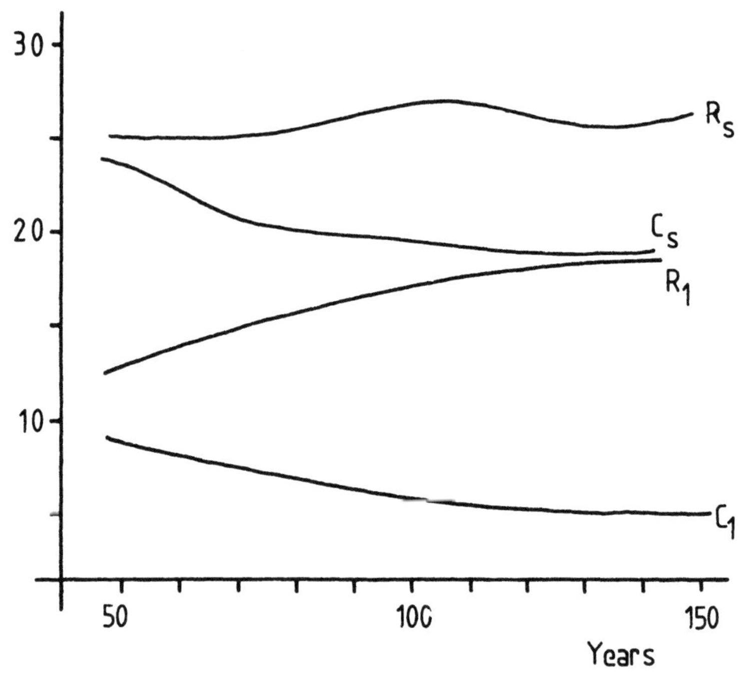

FIG. 8.3 Revenues and Costs of Logging and Sawmilling

Considering equation 8 first, the elements within the brackets show the contribution to overall profit that is made by felling 1 cubic metre of timber and converting it into sawn material, firewood or chips. The various values for the revenues and costs applying to the logging and sawmilling concerns concerned with spruce in the Leningrad area are shown by the appropriately labelled curves in Fig. 8.3. By multiplying the element within the brackets in equation 8 by the mean annual increment of the usable timber assortments, in cubic metres per hectare, a figure is reached for the average, annual profit received by the logging and sawmilling concerns that is sustainable indefinitely. These last factors are shown in Fig. 8.4.

The overall profit, P, reaches a maximum of 13.3 roubles per hectare at the age of 110 years; thus a rotation length of 110 years would be suggested in the Leningrad region's spruce forests. However, at that age the overall output from the forests would be slightly lower than the maximum possible, for the maximum mean annual increment for spruce in this region occurs earlier than 110 years, at the age of 81. Admittedly the difference per hectare would be

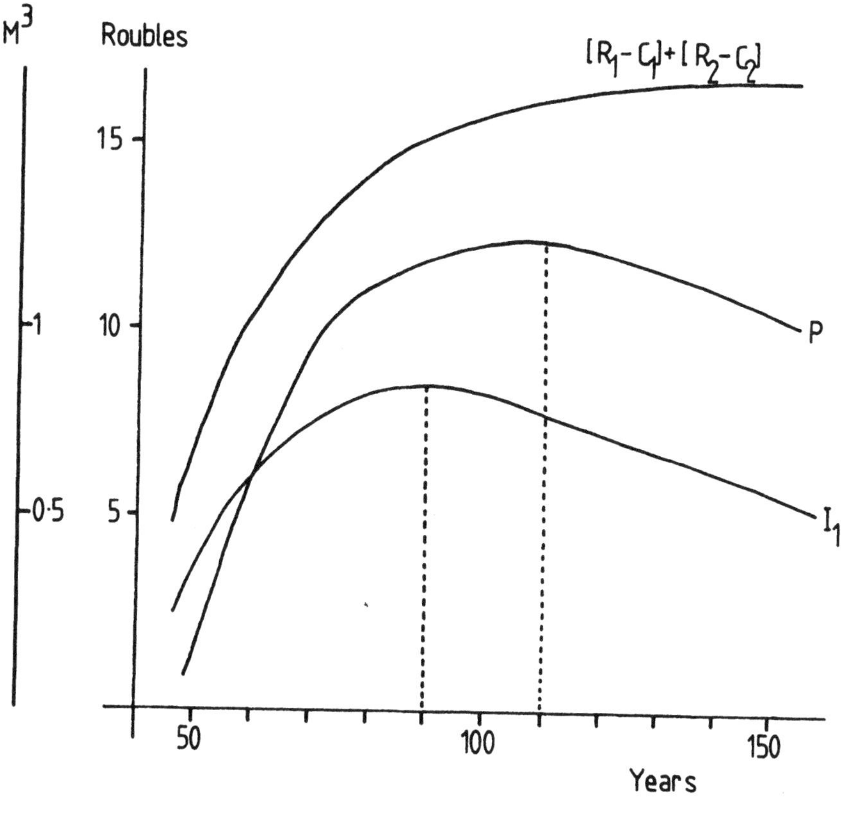

FIG. 8.4 The Optimum Rotation Length

very small, but for large areas of forest the reduction in the overall output might be quite large in absolute terms. To accommodate this, equation 9 is used which adjusts the profit figure, P, to take into account the necessity which would arise to 'import' extra timber from other regions if a rotation age other than that which maximises the mean annual increment is used.

For example, using a rotation length of 110 years leads to an allowable cut of 752.8 thousand cubic metres of spruce per annum from the entire Leningrad district. If the rotation of maximum mean annual increment had been used, i.e. 81 years, this would have allowed 825.6 thousand cubic metres to be logged. The shortfall, only about 9 per cent, is in absolute terms 72.8 thousand cubic metres per annum. Thus, an extra amount of wood above that amount which would anyway have been imported needs to be supplied to the Leningrad region in order to compensate for this. In fact, only 91 per cent of the 72.8 thousand cubic metres needs to be imported as this is the volume of the timber assortments that would have proved to be of use if the 72.8 thousand cubic metres of timber had been felled in the Leningrad region - 91 per cent is the initial conversion ratio for 110 year old spruce timber. Thus, 66.25 thousand cubic metres of sawn timber, etc. needs to be imported. This will involve a transport cost of 4 roubles per cubic metre if imported from the Komi ASSR and the Archangelsk region. Therefore, on a hectare basis, if the area of spruce forest under consideration in the Leningrad region is about 449,000 hectares, the cost per hectare of importing the required timber is 0.59 roubles and the overall, total net profit per hectare, P_t in equation 9, is 12.71 roubles.

Similar calculations may be performed for every age and the effect is to reduce the overall profit level and, possibly, reduce the extension of the rotation length compared with the technical rotation that this methodology tends to produce. In this example, the effect of this adjustment is negligible, and the recommended rotation length remains at 110 years. Table 8.3 shows the resulting rotation lengths arrived at using this formulation, compared to the rotation lengths already recommended on the grounds of maximisation of mean annual increment of the desired assortments. As can be seen (and as would be expected from a consideration of Fig. 8.2) the effect of the *LenNIILKh* criterion is mostly to lengthen the rotation length.

Assessment of the *LenNIILKh* method. This method has some similarities with the existing methods of maximising the mean annual volume increment. It suggests a continuing and constant use of the timber resource at the level of profitability implied by the formulae. It is suitable for use in forests whose first rotation is naturally occurring and costless, in which case the question of discounting the various cash flow streams in perpetuity may be ignored. The conditions when this would hold have been outlined above.

It is a more 'economic' approach to the problem, and takes into account various factors that the purely physical or biological approach to the rotation length ignores. However, disadvantages in the use of prices and costs have been pointed out. The prices for forest products must be the 'correct' ones, reflecting relative scarcities and desirabilities of the various timber assortments before the profit, or indeed any usual economic criterion, may be

TABLE 8.3 Results of the *LenNIILKh* method

	Murmansk area		Karelian ASSR		Leningrad area		Novgorod area		Pskov area	
	(1)	(2)	(1)	(2)	(1)	(2)	(1)	(2)	(1)	(2)
Pine forests with large average tree sizes	170	160	140	121	110	81	110	81	110	110
Pine forests with small average tree sizes	-	-	-	-	120	81	120	81	120	110
Spruce	140	121	130	121	110	81	110	81	90	90
Birch	-	-	85	65	75	55	75	75	75	65

Figures in columns (1) show the figures resulting from the calculations according to the *LenNIILKh* methodology. Columns (2) show the present recommendations.

relied upon to furnish the 'correct' solution.

However, it does explicitly take into account other areas of the forest economy and so represents an improvement on the ordinary criterion. But even so, it is not altogether clear that the adjustment for the extra timber 'imported' into a region is the correct one to make. Indeed, if this criterion were to be imposed on all of the forests of the Soviet Union, it is clear that the overall rotation length would remain unaltered. This is because the adjustment made in equation 9 would ensure that the total usage of timber in each region would remain unaltered and, therefore, the overall rotation length would remain the same. It is not clear if any distributional effects on logging operations would occur if the criterion were given universal applicability.

If, however, this criterion were used in the areas of European Russia, the area where *LenNIILKh* is recommending its use, the effect would be one of lengthening the rotation length there and extending the amount of timber that would have to be obtained from other parts of the country. The criterion would tend to minimise the cost of this operation and so would have much to recommend it, as it would help to shift the weight of logging operations away from the over-logged areas to the more densely forested parts of the Soviet Union at the minimum extra cost.

The Question of Discounting

As noted above, under certain conditions, the technique suggested by *LenNIILKh* is independent of the factor of time and the problem of discounting. The approach by the economists at *LenNIILKh* is to view the forest economy as a whole such that some areas of natural forest are being felled whilst others are being regenerated - which disposes with the need to use discounted cash flow methods as was shown in equations 6 and 7. However, in conversation with the author, some specialists there have dismissed discounting ideas not as being valid but unnecessary in the circumstances, but as being 'mere games with numbers'.

However, the argument about the use of discounting and the effect of time in forestry is quite an important one now being discussed in the Soviet journals. Although the use of discounting principles may not be relevant to most of the Soviet situation, where the new generation of forest replaces already existing stands, there is a use for discounting methods in finding the true value of reproducing a stand of timber instead of some other crop. In this case, the existing stand of timber to be felled would contribute revenue to both reafforestation as well as to, say, an agricultural option. In this case the effect of time on the receipts and costs would have to be taken into account explicitly by some sort of discounting technique.

Some Soviet specialists are arguing that discounting should be used in determining the rotation length. Others argue, not as I have done earlier, that discounting may not alter the result and so may be ignored, but that discounting is conceptually wrong even for new plantations of trees. Some authors, for example Tsimek (1975) and Yanushenko (1968), consider the use of discounting wrong. Such specialists consider the long growing period of a forest to be 'part of the material, technical side of forestry and not an economic factor' (Voronin and Panitseva, 1976). Also, the influential Anuchin (1977a and b) recommends physical criteria for the definition of the rotation length and gives over only one paragraph in his book to the economic maturity of a forest. There he states that the optimum rotation length might be that which maximises the output of the most valuable assortments of timber, although 'calculations show that the age of technical and economic maturity lie close to each other'.

Chuprov (1975) follows a similar line to that taken by the specialists at *LenNIILKh* and derives rotation lengths on the basis of profit per hectare, but only considers the profit from logging. He defines his criterion as the maximisation of P defined by the equation below.

$$P = \frac{R_1 - C_1}{T}$$

Here P, R_1 and C_1 have the same meaning as earlier but T refers to the rotation length. Like the *LenNIILKh* team, he recommends slightly longer rotation lengths to those in use at the moment.

The problem that Soviet economists face if they argue for the use of discounting methods is the fact that discounting over a life span of a forest brings about a very substantial reduction in the value of future revenues. Thus, at the 8 per cent norm, 1 rouble in fifty years time would be valued at only slightly more than 2 copecks. Whilst there seems to be a trend towards the acceptance of the idea of discounting techniques in forestry (as a reflection of their more general use in the economy as a whole) this effect of discounting over the time periods involved in forestry is presenting a stumbling block to those advocating its use, as indeed it did to those who advocated its use in the West many years ago.

Two approaches to this problem are apparent. One is the suggestion that simple interest calculations should be used in the place of the usual compounding formulae. But the most common approach is the use of a lower discounting norm, usually between 2 and 3 per cent. One reason for taking such a low figure comes from the work of Goffman (1974) who showed that if the discount rate lies between 2 and 3 per cent the following inequality is true.

$$\frac{q(T).p(T)}{r.T} > q(t).p(t) + \frac{q(T).p(T)}{(1+r)T - 1}$$

Here q(T) is the volume per hectare at age T, p(T) is the price of 1 cubic metre of timber at age T and r is the discount rate.

Here the left-hand side of the inequality shows the value, discounted in perpetuity, of a stream of receipts from cutting the forest by an equal volume per annum through time. The right-hand side shows the value of felling every tree now and having the entire forest stock coming to maturity in T. years. That is, the inequality shows those conditions when the continual and uninterrupted use of the forest will be more economical than the use of the forest in an interrupted manner. Of course, Goffman ignores the problems that would be created with such a peculiar pattern of forest use, but concludes that the discount rate *should* be that value which allows the continual and uninterrupted use of the forests. Thus, it should be between 2 and 3 per cent.

Finally, work by Seperovich (1976) shows the effect of variations in the discount rate on the economics of oak plantations in the central European part of Russia. He uses a 'coefficient of economic effectiveness' which is the net discounted revenue divided by discounted costs. He assumes no standing crop, the first revenues accruing at the end of the rotation. As a result he comes to a series of conclusions that sound remarkably like the conclusions that might be reached by forest economists working in the West. He concludes that long rotations are uneconomic and that;

> '... with the calculation of the economic effectiveness of forestry a discount rate of about 2 per cent should be used by the industry in calculations involving periods of up to 100 years.'

Conclusions and Implications

The definition of the rotation length, when applied to the volume of accessible timber stocks, gives the maximum volume of timber that can be supplied on a sustainable basis to the Soviet economy. At present the research being undertaken to define better the rotation length may be rather crudely divided into two basic approaches to the problem.

On the one hand are those who use discounted cash flow techniques and arrive at the wrong answers for the right reasons. And on the other hand are those who eschew discounting methods over the time periods involved in forestry as methodologically wrong, and so reach correct solutions for the wrong reasons.

There appears to be a good case for the extension of the rotation length so as to allow economies in the harvesting and timber processing industries - see Figure 8.2. But calculations in this method need to be made so that the total effect of such an extension are seen. For example, what would be the true cost of using longer rotation lengths if this meant that more remote areas of forests would have to be brought into production?

Whether discounting methods become finally acceptable for such long periods of time as the rotation length of a forest depends upon whether the move to their increased use in the economy as a whole is sufficient to give enough drive to their advocates in forest economics. This would seem a likely eventual outcome. Already, discounting techniques are accepted by nearly all Soviet economists for short periods of time. For example, Yanushenko, who does not approve of their use for the determination of the rotation length, does include a section in his book on their use in the more normal capital appraisal situation. If they do find universal use, however, it is to be hoped that they are used with great care so that the calculations are formulated taking into account the crop of standing timber. This will allow a correct definition of the rotation length and not an artificially shortened one as obtained by Seperovich (1976). However, there might be some attractiveness in a methodology that provided an apparently valid reason for the reduction of the rotation length. In shortening the rotation the allowable cut is increased in the short run - using the term short run in a forestry sense. This would offer a short-term method (say for the next thirty years) of increasing the output of the forest industries, although, of course such a path would only serve to accentuate the problems of expanding the output of timber to the economy in the long-term as the average age of the accessible forest fell and the average annual volume increment fell below the maximum possible.

If such techniques do not achieve universal acceptance, although the definition of the rotation length may still be arrived at in a satisfactory manner, errors are likely to be made in other areas. Although the correct rotation length for a given species may be arrived at without using discounted cash flow methods, given the assumptions arrived at earlier, the choice of species cannot. A simple example illustrates this point. In Yanushenko's book (1977) he presents his framework for the decision of what species to plant. His criterion is defined as K in the equation below.

$$K = \frac{\text{Revenues - Costs}}{\text{Costs}}$$

He presents figures for areas in White Russia with pine and birch forests and shows that for pine the value for his criterion, K, is 2.28 whilst that for birch is only 0.9. He would conclude here that pine should be allowed to regenerate at the expense of birch when the forest is logged over.

However, this might be the wrong conclusion. Assume a rotation length of 100 years for pine and only 50 for birch; in this case account has to be taken that two crops of birch can be harvested every 100 years to only 1 of pine. This Yanushenko does by using a 100 year accounting period and defining the costs and revenues as 100x/t, where x is the cost or revenue of 1 rotation and t is the rotation length in years. But the fact that, for the pine and birch crops, the pattern of the cash flows will be different must be accounted for.

If the definition of his K is manipulated the ratios of revenues to costs may be arrived at - it will be equal to the coefficient plus 1. If the costs for both species are assumed to be equal to 1 unit occurring at the beginning of the rotation, the cash flow position for both species can be arrived at. These are shown in Table 8.4. Using a discount rate of 2 per cent leads to a net discounted revenue in perpetuity of -0.47 for birch and -0.62 for pine. Thus although both species would appear to be ineffective, birch would be the better species of the two. This result might be expected because, although its return is less in absolute terms, it reaches maturity earlier, which is an important factor when discounted cash flow methods are to be used.

Therefore, although the discount rate argument may be being fought around the question of the rotation length (where it may be irrelevant) its true relevance and the real need for it lie elsewhere.

TABLE 8.4 The Choice of Species

		Now	+50	Year +100	+150	+200
Pine	Cost	1	-	1.00	-	1.00
	Revenue	0	-	3.38	-	3.38
	NDR	-1	-	0.33	-	0.05
Birch	Cost	1	1.00	1.00	1.00	1.00
	Revenue	0	1.90	1.90	1.90	1.90
	NDR	-1	0.33	0.13	0.05	0.02

9
Possible Responses to the Timber Shortage

INTRODUCTION

If the prognoses in the earlier parts of this book are correct the output of timber from the Soviet Union's forests is unlikely to grow rapidly in the future. This will have its ramifications in other sectors of the country's economy as timber is an important input into almost every other branch of industry. There will be pressure, therefore, to implement strategies to increase the amount of usable timber supplied to the domestic economy without the even greater over-cutting of the forests already being harvested.

In this chapter three such strategies will be examined. The first is the more intensive use of the Group I forests; these are the forests which fulfil some protective role and are, as yet, under-utilised. The second strategy to be examined is the increased use of the timber growing on a particular site. Already the fact that many of the broadleaved trees are left standing after harvesting operations has been noted but there are also moves to increase the use of branches, logging waste and even tree roots for the production of chips for industrial use. Finally, the possibility of diverting exported timber to the domestic economy will be considered.

GROUP I FORESTS: THEIR STRUCTURE AND POTENTIALITIES

Introduction

Group I forests are those forests which fulfil a protective role of some description, or those which form green belts around urban centres. Such forests mostly come under the control of *Gosleskhoz* and its regional subcommittees, although some of the forests in collective and State farms come under the jurisdiction of the agricultural authorities.

The management of these forests is much less intensive, in the sense of volume production, than that of the forests of the other groups. The management objective of Group I forests is not the maximisation of possible timber output but the 'production' of some degree of protection for a slope or agricultural area, or the provision of some amenity. Thus, rotation lengths tend to be longer than those in other groups of forests, clear-felling is used much less often and selective cuttings are more common. However, as such forests constitute about 16 per cent of the total forest area of the Soviet Union, their potential output is obviously of significance for the Soviet Union's forest industries.

According to the 1973 survey the total area of Group I forests was 194.3 million hectares. The overall figure in 1978 was 208.6 million hectares of forest of which 115.4 million was classified as being under forest cover. The increase in area came about because of the transfer of some forests from Groups II and III to Group I, especially in areas around Lake Baikal, in Kazakhstan, in Latvia and in the Moscow region (Nikolauk, 1979).

Unfortunately, there seems to be no general breakdown of the forests of Group I for a later date than 1973. Referring to these figures, of the 194.3 million hectares of forest, 60.3 million hectares, or about 31 per cent, was located in the European part of the Soviet Union. The coniferous species accounted for 63.6 per cent of the forests although in the European part of the country this figure was only 52.7 per cent. The volume of the major species in the mature and over-mature categories is in excess of 6,000 million cubic metres, or approximately 30 cubic metres per hectare. This low figure is because of the fact that the area of Group I forests refers to the total area, including unforested area included in the Group I category.

From these figures it is obvious that the annual output of timber from Group I forests is capable of quite a significant increase from the present figure of about 22 million cubic metres per annum with probably no appreciable effect on the protective or amenity properties of those forests. Also, a significant proportion of these forests is located in the European areas of the Soviet Union, exactly in those areas where the deficit in timber supply is being most acutely felt. Thus, many authors consider that the forests in Group I have to be better managed if the expansion of output from Soviet forests is to be maintained or accelerated.

A Brief History of Protective Forestry

The necessity to define protective forestry and to impose special laws on their use has long been recognised in Russia. In 1701, 1716 and 1719 Peter the Great introduced edicts that limited the clearing of forests within 50 versts (53.5 kilometres) of large rivers and 20 versts (21.4 kilometres) of small ones. Whilst it is usually claimed that this was designed to be a protective measure, there is the strong possibility that this legislation was prompted by the same reasons that similar laws were introduced by Elizabeth I in England. This was to stop charcoal burners from felling trees near to navigable rivers as this represented the only accessible source of timber for ship-building. More of Peter the Great's legislation in 1723 suggested the careful management of forests near the sea, around towns and along roads was being taken more seriously.

However, as the industrialisation process in Tsarist Russia slowly progressed, so the demand for wood increased. The iron smelting industry in the Urals accounted for much of the timber felled in that region and many concessions along rivers and near to large ports were sold to foreign firms for exploitation. It was common practice for land-owning nobility to sell its forests if short of money. In 1888 the first actual forest law was passed which defined protective forests along rivers, hill-sides, valleys in areas susceptible to erosion and timber stands that prevented the movement of sand dunes. However, the area actually covered by such laws was quite limited. The provisions of the law applied only to the European parts of Russia and the Caucasus, although in 1904 the regulations were extended slightly. The law, ordered it is said on the initiative of Tsar Alexander III, was intended primarily to stop excessive clearing of forests and prevent the conversion of forest land to other uses. Approved working plans were required in the areas under the control of the law and any cutting not contained in the plans or in excess of the planned amount incurred a penalty. Whilst the law was, apparently, intended to apply to the nobility as well as the peasantry it was mainly aimed at the latter; this was partly because the peasant holdings had been rapidly depleted in the years before the law, either directly by the peasants themselves, or by their selling their timber holdings to timber merchants. Also, of course, the conservative Tsarist Government saw such conservation as stabilising the peasant communities.

After the revolution and Bolshevik take-over, the forests were placed under central control. On 27 May 1918 a decree entitled 'On Forests' divided the forests of the RSFSR into two basic groups. These were the exploitable forests which could be used for 'any timber or monetary benefit' and protective forests the use of which was supposed to be limited. However, circumstances stopped these intentions from being put into practice. During the civil war that followed the revolution the easily accessible forests were officially released for clearing to provide fuel. Also, forests within 50 kilometres of Leningrad and 30 kilometres of Moscow were released from control to provide the cities' inhabitants with fuel. In 1923 the Forest Code of the RSFSR was drawn up; this was amended and appeared in its final form in 1929. This Code expanded the definition of protective forests to include not only watershed management but also forests used in the control of drifting snow near roads and railways, those used for shelter belts in the agricultural context and those forests of some scientific interest. However, it seems that, again because of a lack of strict control, these codes had only limited success.

Another attempt to control the amount of cutting in the easily accessible forests in areas where watershed management was significant was made on 2 July 1936. A resolution was passed by the Soviet of Peoples' Commissars entitled 'On the Formation and the Administration of Protective Forestry and Forest Plantations of the Soviet Peoples' Commissars and on the Division of Watershed Areas'. Seventy million hectares of forests were brought under the Administration's jurisdiction in the basins of the Dnieper, Don, Volga and in the Urals. In the immediate vicinity of the rivers themselves cutting was restricted to sanitation cuttings, whilst in the other areas cutting was restricted to the mean annual increment. But at the beginning of the war unrestricted felling was allowed, but only as a 'temporary measure'. In 1943 the legislation was drawn up to rectify this position, so it was not until 23 April 1943 that the present system of dividing forests into different groups for the purposes of management was introduced. Resolution number 430 of the Soviet Peoples' Commissars of the USSR - 'On the Procedures of Allocation of Allowable Cut in the State Forests of the Soviet Union' - was passed to rationalise the use of various forests in the different regions of the Soviet Union. The three groups of forests, similar in definition to the groupings used today, were formed under this resolution. The three groups are described in Chapter 2.

Group I forests originally included the State forest reserves, soil and field protecting forests, forests of the green belts in the vicinity of towns and industrial areas, and firewood forests in the Steppes. Later the forests in the watershed areas listed earlier were included as were the forest belts along the railways and major roads which were designed to protect them from drifting snow. In these forests, at first, clear-felling was forbidden and only thinnings of one type or other were allowed.

In 1943 the actual area of Group I forests was relatively small, although it grew quite rapidly as more forests with a watershed management role and pretundra forests were included in the group. The latter group of forests were located in a band adjoining the tundra area in the north of the Soviet Union; they were to be left to provide a continental scale 'wind break' to reduce the effect of north winds on the climate of the interior of the country. The changes in the area of forests included in Group I are shown in Table 9.1.

TABLE 9.1 Changes in the Area of Group I Forests (millions of hectares)

	1956	1961	1966	1973
Green belts	9.9	13.1	14.6	18.4
Forests along rivers, around lakes, etc.	29.1	53.7	55.5	71.3
Nut and seed producing forests	2.0	6.2	9.0	10.0
Pretundra forests	0.0	72.6	55.2	52.9
Other	21.9	24.5	27.0	41.7
TOTAL	62.9	170.1	161.3	194.3

Compiled from various sources.

In 1953 a more codified approach to the management of Group I forests was adopted. Felling based on sylvicultural criteria was allowed. Rotation lengths equal to the sylvicultural maturity of the various species were outlined. This was partly as a result of the low level of output that was originating in these forests after the introduction of the 1943 regulations and partly as an attempt to manage these forests better in their various protective and amenity roles. In 1953 1.5 million cubic metres of timber were extracted from Group I forests, which constituted only 0.05 per cent of their mature and over-mature growing stock. In 1954 this had increased to 8.2 million. However, it was still felt that the actual volume felled in Group I forests was too small compared to the productive potential of these forests. So in 1967 *Gosleskhoz* worked out 'The Methodology of Calculating the Allowable Cut in Forests of the Forest Fund of the Soviet Union'. According to these instructions, it was possible to include a greater area of the forests included in Group I in the area that acted as the basis for calculating the allowable cut. In the category of forests that could be exploited more intensively were some of the forests along rivers, beside lakes, protective belts along railways and main roads, some forests which had a field and soil protecting role and some of the forests making up green belts around urban areas.

According to *Gosleskhoz's* figures for 1971, of the 343 million cubic metres of timber felled in the main felling operations in the Soviet Union's forests, about 20 million, or about 6 per cent, were the result of felling in Group I forests. Table 9.2 shows the growth of output of Group I forests over the 1960s; as can be seen, most of the volume felled in Group I forests has come from operations in European Russia. This is clearly a bonus to the forest industries as it is the European part of the country that most feels the deficit of timber. Also, it is evident from this table that the regulations of 1967 did not radically alter the level of exploitation of these forests. Indeed, in recent years the level of output has increased only slightly to about 22 million cubic metres per annum.

TABLE 9.2 The Ouput from the Forests of Group I (millions of cubic metres)

	1960	1965	1966	1968	1970
Total volume	14.2	19.7	19.9	20.3	19.5
Output in the European part of the Soviet Union	12.5	17.4	17.5	17.3	16.7

Source: Nikolaenko *et al.*, 1973

Group I Forests at Present

Table 9.3 shows the distribution of Group I forests in the Soviet Union; the figures are the results of the last complete survey which was carried out in 1973. All of the figures in this table refer to actual forest covered area. The total land area designated as Group I forest is larger. 194.3 million hectares of land was considered to be the area of Group I forests in 1973, in Table 9.3 it is evident that only 94.735 million hectares are actually under forest cover. The rest is made up of roads, railways, rivers, lakes and areas under scrub.

TABLE 9.3 Distribution and Uses of the Group I Forests

(All figures are in thousands of hectares)

	Total area	Green belts	Belts along rivers and roads
TOTAL	94,735	11,187	36,454
European part	38,428	8,442	13,262
RSFSR	84,569	9,062	34,726
European part	30,961	6,451	11,594
Regions			
North-west	14,835	1,502	4,420
Central	4,263	1,531	2,496
Volgo-vyatka	2,011	463	1,402
Central Black earth	703	229	173
Povolzhski	3,089	753	1,387
North Caucasus	1,725	238	107
Urals	4,248	1,685	1,606
Western Siberia	8,274	641	2,212
Eastern Siberia	25,734	1,036	7,294
Far East	19,687	984	13,639
Ukraine	1,906	840	336
White Russia	1,280	429	849
Georgia	2,305	161	27
Azerbaijan	747	11	4
Lithuania	474	245	142
Moldavia	142	44	10
Latvia	414	141	228
Armenia	253	16	23
Estonia	216	104	49
Kazakstan	1,742	114	57
Uzbekistan	200	5	-
Kirghizia	412	14	-
Tadzhikistan	284	1	-
Turkmenistan	61	-	-

Source: Vorobiev *et al.*, 1979

Soil protection	Amenity forests	Pretundra	Other	Group I forests' percentage area*
9,184	580	26,468	10,862	14
5,237	475	8,720	2,292	26
4,922	226	26,468	9,165	13
3,188	121	8,720	887	24
7	11	8,720	184	20
188	32	-	16	35
111	11	-	24	20
169	-	-	132	66
934	4	-	11	36
1,312	55	-	13	62
449	8	-	500	18
1,332	20	2,990	1,079	11
379	74	10,913	6,048	12
41	11	3,845	1,150	9
575	47	-	108	36
-	2	-	-	27
703	263	-	881	98
543	4	-	185	99
-	9	-	78	40
67	-	-	21	70
26	5	-	14	26
86	15	-	113	100
49	9	-	5	25
1,433	-	-	138	22
186	-	-	9	25
263	-	-	135	100
273	-	-	10	100
58	-	-	-	50

* shows the area of Group I forests as a percentage of all forests in the area.

About 90 per cent of these forests are located in the RSFSR. Although on a percentage basis the Group I forests in the European part of the RSFSR represent a greater proportion of the entire forest area, in absolute terms Group I forests in Siberia and the Far East cover a far greater area. In some Soviet Republics, Armenia, the Kirghiz Republic and Tadzhikistan, all of the forests are Group I. In other Republics such as Georgia and Azerbaijan Group I forests constitute by far the largest category.

In terms of their designation, Group I forests having a watershed management role along rivers and pretundra forests constitute the largest element, making up about two-thirds of all Group I forests.

As was mentioned before, some of these forests are included in the 'Rules Concerning the Cutting of Plains Forests in the European Part of the RSFSR', and *Gosleskhoz's* 'Methodology' of 1967. All of Group I forests may be included in the area of State forests used in calculating the allowable cut except the following: certain areas of green belts, field and soil protection forests within 100 metres of unforested land, protective belts along the side of roads and railways, forest on steep slopes, forest belts along rivers, canals and lakes, small forest areas of less than 100 hectares situated in unforested areas and pretundra forests. Given these restrictions, the ability of Group I forests to supply timber to the Soviet economy is limited. Table 9.4 shows the break-down of the forest covered land in the European part of the Soviet Union that is included in the area used to calculate the allowable cut. The figures refer not only to the European part of the RSFSR but also to the other Soviet Republics in Europe. The figures in Table 9.4 differ slightly from the sum of the areas of Group I forests in Europe found in Table 9.3

TABLE 9.4 The Allowable Cut and Group I Forests

	Actual Forest Area at 1 Jan 1970	
	Total	Included in the allowable cut
Green belts	9,339.0	5,747.1
Belts along rivers, lakes and other water areas	11,096.0	7,800.5
Field and soil protection	6,097.9	3,137.4
Belts along roads and railways	2,022.8	1,624.8
Seed and nut orchards	324.4	254.4
Pretundra	8,700.6	0.0
Other categories	2,019.9	732.1
TOTAL	39,600.6	19,296.4

All figures are in thousands of hectares and refer only to the European part of the Soviet Union.

Source: Nikolaenko *et al.*, 1973.

because the figures in Table 9.4 refer to the survey made in January 1970.

As may be seen from the table, of the 39.6 million hectares of Group I forests in the European part of the Soviet Union, about half 19.296 million hectares are taken into account in calculating the allowable cut. From this 19 or so million hectares comes most of the timber that is harvested in Europe's Group I forests. Before we can assess whether this figure represents over- or under-exploitation of the growing stock, and therefore whether there exists any potential for expansion of output from these forests, we need to know more than the mere areal characteristics of these forests.

Species Composition and Age Structure of Group I Forests

According to the 1973 survey the total growing stock of Group I forests was 10,020 million cubic metres. On an area basis, about 64 per cent of Group I forests were made up of the coniferous species. The predominant coniferous species is larch, which accounts for 21 per cent of the total area; following this are pine, taking up 19 per cent, spruce, 14 per cent and cedar, 9 per cent. The remaining 37 per cent of Group I forests are made up of broadleaved species. Birch predominates here, taking up 16 per cent of the total area and oak accounts for 5 per cent.

In the European part of the Soviet Union the figures are slightly different to the nationwide averages. There is, of course, much less larch. But more importantly, the area of Group I forests in the European part of the country taken up by the more valuable coniferous species is less, as a proportion, than it is in the country as a whole. Coniferous species account for only a little under 53 per cent. About 30 per cent of the total area is pine and much of the remainder spruce. Of the area of Group I forests taken up by broadleaved species, the predominant species is birch. So the value of the European Group I forests to the Soviet economy is less than would appear from a consideration of the national figures for the species composition of this group.

A similar position is reflected in figures showing the age structure of Group I forests. For the country as a whole almost one-half of Group I forests are classified as being 'mature' or 'over-mature'. This would suggest that there is some latitude for a more liberal felling regime without loss of the protective or amenity attributes of these forests. However, 70 per cent of the mature and over-mature timber is situated in Siberia and the Far East where ample alternative sources of timber exist. In the European area of the Soviet Union the position is different; here Group I forests are much younger than they are in the country as a whole. Only 22 per cent of the European forests are in the 'mature' and 'over-mature' categories. The reason for this, of course, is that the intensity of logging in the European parts of the Soviet Union in the past has produced a situation where forests in general are much younger than they are in the country overall.

There are some figures that illustrate this position and these are presented in Table 9.5. These figures come from the 1970 survey and so do not easily tally with the figures in Table 9.3. The figures show that the younger age groups predominate in the European

TABLE 9.5 Age and Species Composition of European Group I Forests

	TOTAL	Middle aged stands	Stands almost mature	Mature and over-mature stands	Mature and over-mature growing stock
Coniferous species	15,204.4	4,930.9	1,575.7	3,495.2	640.84
inc. pine	9,826.6	3,174.1	853.8	1,321.7	212.15
spruce	4,661.6	1,515.9	617.1	1,840.6	314.94
Broadleaved species	15,695.6	6,447.7	2,118.4	3,315.3	602.40
TOTAL	30,900.0	11,378.6	3,694.1	6,810.5	1,243.64

All figures are in thousands of hectares except for the column headed 'over-mature growing stock' which is in millions of cubic metres. Figures exclude pretundra forests.

Source: Nikolaenko *et al.*, 1973.

Group I forests. This is most marked in the coniferous species where there is a noticeable lack of 'almost mature stands'. Whilst this will not affect the output from these forests if they are to be regarded as a short-term source of timber for the Soviet economy, a stop-gap until the investment plans for the Group II and III forests come into fruition, it will be of significance if they are to be considered as a viable long-term source of timber.

An Assessment of the Potential Output of Group I Forests

From the figures in Tables 9.4 and 9.5 it is possible to arrive at an estimate of the potential output of timber from Group I forests. From Table 9.4 it is apparent that, ignoring the pretundra forests which are excluded from Table 9.5 also, about 64 per cent of all Group I forests in the European part of the country can be included in the allowable cut. Assuming that this figure can be applied to all age classes and all species, Table 9.6 can be constructed. In this table the first rows referring to both coniferous and broadleaved timber are reproduced from Table 9.5. The second rows show the 64 per cent of those areas and volumes in the first rows which it is assumed may be included in the allowable cut, and the third rows show the areas and volumes not included in the allowable cut, that is the differences between the first and second rows. Remembering that the verbal descriptions of maturity used in the Soviet Union correspond exactly to age classes rather than to less well defined sylvicultural conditions, the figures in Table 9.6 can be interpreted in terms of age classes.

Taking coniferous timber first, these species are divided into 20 year age groups. 'Young stands' are stands between the ages of 1 and 40 years, 'middle aged stands' have ages between 41 and 60 years,

TABLE 9.6 Potential Output from Group I Forests

	Stands almost mature	Mature and over-mature stands	Growing stock of mature and over-mature stands
Coniferous species			
TOTAL	1,575.7	3,495.5	680.84
included in the allowable cut	977.0	2,167.0	397.00
not included in the allowable cut	598.7	1,328.2	243.84
Broadleaved species			
TOTAL	2,118.4	3,315.3	602.40
included in the allowable cut	1,313.0	2,055.0	374.00
not included in the allowable cut	805.4	1,260.3	228.4

Figures are in thousands of hectares and millions of cubic metres.

Source: see text.

'almost mature stands' have ages between 61 and 80 and stands older than 81 are classified as 'mature and over-mature'. The normal rotation length for coniferous trees in the European parts of the Soviet Union is in the region of 80 years. Assuming one extra age class for Group I forests that can be included in the allowable cut and two extra age classes for those which are not included gives rotations of 100 and 120 years respectively. If 397 million cubic metres of timber are to be found in the 'mature and over-mature' age classes of coniferous forests that can be included in the allowable cut, and if these forests are to have a rotation length of 100 years, it follows that this volume of timber can be felled over a 20 year period. This gives an annual output of 19.85 million cubic metres - ignoring the annual increment of timber during this 20 year period. If the 243.84 million cubic metres of timber in the Group I forests not included in the allowable cut has a rotation length of 120 years, this volume may be felled over a 40 year period and this will give an annual output of 6.096 million cubic metres. This gives a total annual output of coniferous timber from the Group I forests in Europe of 25.946 million cubic metres over a 20 year period.

Turning now to the broadleaved species, a similar set of calculations can be performed. The same verbal descriptions of stand maturity which refer to specific 20 year age classes when applied to coniferous species are applied to only 10 year age classes in the context of broadleaved species. 'Young stands' consist of trees between the ages of 1 and 20 years, 'middle aged stands' are aged between 21 and 30, 'almost mature stands' contain trees between 31 and 40, and 'mature and over-mature' timber is older than 40 years. A common rotation length for broadleaved species in the European parts of the country is between 41 and 60, that is, anywhere between the fifth and sixth age classes. Assuming that the suitable rotation length for broadleaved Group I forests that can be included in the allowable cut is 50 years, and 60 for the same forests that are not included in the allowable cut, the output of timber from the broadleaved Group I forests over the years 1980 to 2000 can be calculated.

Over the ten years from 1980 to 1990, all of the 'over-mature' and 'mature' timber in the broadleaved forests that can be included in the allowable cut will be felled, giving 37.4 million cubic metres per annum - ignoring the increment during that period. From the forests not included in the allowable cut 11.4 million cubic metres per annum will be available, as the 'mature and over-mature' age classes will be felled over a period of 20 years. This gives a total output from these forests of 49.8 million cubic metres per annum.

During the period 1990 to 2000 the forests not included in the allowable cut in the age class 41 to 50 will cover an area of 1,313 thousand hectares, this being the area of the 'almost mature' age class at present. Assuming that, on maturity, this has the same stocking as the present 'mature and over-mature' forests, this will yield 23.9 million cubic metres over the ten years in question. Thus, between 1990 and the year 2000 an output of 34.3 million cubic metres per annum is possible.

Conclusion

Combining the coniferous and broadleaved predictions, it would seem possible that 65.746 million cubic metres of timber could be harvested from these forests between 1980 and 1990 and 60.246 million cubic metres of timber from 1990 to 2000. In both of these ten year periods it would be the broadleaved species that would make up the bulk of the output. These estimates are probably the maximum volumes that could be harvested from these European Group I forests as the assumptions about the proper rotation lengths have been fairly radical and it is likely that longer rotations would be insisted upon than the 100 and 120 years for coniferous species and 50 to 60 years for broadleaved stands. However, the figures are greater than the present output of Group I forests and so it is probably safe to conclude that sylviculturally the output from these European forests can be increased by quite a large amount.

But this conclusion assumes that other resources are in good supply. The technology and manpower over and above that already required in the other forests in the European parts of the country must be available to take advantage of the available timber resource. This is clearly not the case as it has already been seen that the labour supply is an important constraint to any increase in output of the Soviet Union's forests. This is true in both Siberia and in Europe. The only real chance of taking advantage of this extra potential output is if workers who previously did not work in forestry can be attracted into work in the Group I forests in the European parts of the country maybe for reasons of their more favourable location close to urban centres.

As well as this problem, it was seen in Chapter 3 that the broadleaved allowable cut in the European part of the Soviet Union is under-utilised. It is only in the case of the coniferous species that the allowable cut was being exceeded. In this case the contribution that the broadleaved species in the Group I forests can make to the economy as a whole is only marginal. Thus, the output from these forests that it might be worth exploiting is the coniferous potential, an output of about 28 million cubic metres per annum.

From this analysis of the potential of these forests, therefore, it would seem that there is no significant supply of timber to the economy that would alleviate the problems in the other areas of the logging industry.

LOW QUALITY WOOD

Not all of the timber growing on a site goes for consumption. Some is left standing, as has been discussed, but more timber (firewood, branches and tree tops and root systems) remains unused by the timber industries. There is increasing pressure for its use in the future.

Firewood

One source of additional timber could be the use of wood which for the present, goes as firewood. A little less than 20 per cent of the standing timber which is felled is of firewood 'quality' and converting it into more valuable chips is a way of increasing the volume of 'industrial' or 'useful' wood produced by the *lespromkhoz* and at the same time increasing the value of sales and the level of profitability, all important indicators in plan fulfilment.

Table 9.7 shows the changes in the amount of firewood harvested in the Soviet Union since 1913. There are two sources of firewood for the economy identified in the table. First is the volume of industrial logging output and the proportion of that output which goes for firewood; these figures are taken in part from the data presented in Table 1.1 and refer to the logging industry proper. The other columns in Table 9.7 refer to the non-industrial output of timber, the bulk of which is felled by the agricultural sector in the loosest sense of the term, that is, by the workers in *kolkhozi* and *sovkhozi* and by the rural population.

As would be expected, the importance of firewood has decreased markedly since 1913. In the early parts of this century firewood was the major forest product, and even at the outbreak of the Second World War it accounted for much more than half of all the timber harvested in the Soviet Union by the industrial and non-industrial sectors.

Now firewood only accounts for about 20 per cent of the total timber extraction, although in the non-industrial sector this figure is still in excess of 80 per cent. Clearly, if it were possible to transfer this to the area of wood processing a very real increase in output would be feasible. However, it is unlikely that such a

TABLE 9.7 Firewood in the Soviet Economy

	Volume of industrial logging	Of which firewood		Volume of non-industrial logging	Of which firewood	
		M^3	%		M^3	%
1913	62.0	24.1	55.0	232	204	87.9
1922	57.8	30.5	52.8	269	209	77.7
1928	61.7	25.9	41.9	242	215	88.8
1932	164.7	56.0	34.0	223	201	90.1
1940	246.1	128.0	52.0	136	130	95.6
1945	168.4	107.3	63.7	140	130	92.9
1950	266.0	133.3	50.1	100	92	92.0
1955	333.9	121.9	36.5	70	63	90.0
1960	369.5	106.5	28.8	40	30	75.0
1965	378.9	103.1	27.2	38	32	84.2
1970	385.0	84.7	22.0	37	30	81.1
1975	395.0	70.3	17.7	35	29	82.6

Figures in millions of cubic metres.

Source: Lobovikov and Petrov, 1976 and Table 1.1

transfer is possible, for firewood is an important, although small, source of fuel to the Soviet economy - especially for domestic heating in rural areas. At the present moment about 1.3 per cent of the fuel needs of the Soviet Union is supplied by firewood, and if the performance of the rest of the fuel industry is lacking in any way, there will be pressure not to reduce the supply of firewood to the economy.

Already these pressures are apparent. There are many articles in the literature calling for the savings of fuel in the operation of logging equipment and also for the more efficient supply of firewood to the community. For example, A.I. Korablev (1980) from the Ministry for the Fuel Industry says;

> 'The last winter was considerably better than the winter before, but the shortage in supplies of fuel, including firewood, has not been eradicated. In many rural areas and even in small towns it is usual to see queues for firewood ...'

Korablev lists a number of reasons for this state of affairs. First, he bluntly states that the logging enterprises are simply not fulfilling their plans with regard to the supply of firewood. Secondly, he argues that the lack of supplies of firewood are, in part, the result of a lack of emphasis on a more complete forest management policy. By ignoring such operations as thinnings, especially in the Group I and II forests, a useful supply of timber suitable for fuel is lost. Finally, Korablev suggests that the pricing system is also partly to blame. In an argument that recalls what has been said in Chapter 7 he says;

> 'The point is that, in order to stimulate a larger output of industrial timber (as opposed to firewood), the price of industrial timber assortments was set higher than the average level of costs, but the price of firewood was set considerably lower than actual costs. It is evident that some planners and members of the committee for prices consider firewood to be a waste from the timber production process and not a fuel source.'

The Ministry for the Fuel Industry itself has a forestry division that works in a similar way to the enterprises of *Minlesbumprom*. It is also undertaking an investment programme similar to that of the main ministry to increase the level of mechanisation and the productivity of labour, although figures suggest that the present level of mechanisation is below that of the average for *Minlesbumprom*. As well as firewood these enterprises produce 'by-products' such as sawn timber and furniture.

Although the costs per calorie of heat is higher if firewood is used compared to other fuels such as coal, oil or electricity, it remains an important fuel source in certain areas. So there appears to be a need to rebalance the price received for firewood compared to the price received for low quality wood going for, say, chip production. This would be, in part, a reversal of what occurred in the 1966 price reforms but it would seem that any move on the part of the timber industries to increase their supplies of raw material at the expense of firewood are likely to meet stiff resistance.

Stumpwood

One area of potential supply of cellulose is stumpwood - that is the volume of wood left in the ground after the tree has been felled. As the emphasis on full tree extraction increases the tendency is for some authors to talk about total biomass extraction and utilisation. They point out that it is irrational to assess the volume of available material in terms of the volume of the tree trunk alone and they want to see branches, roots and leaves even included in the idea of the yield from a particular site. Tauber (1979) considers that it will not be long before a total biomass indicator replaces the present simple indicator of 'cubic metres of timber' that the plans contain.

The removal of stumps on a commercial basis in the Soviet Union is limited to the removal of pine stumps for the chemical industry. However, experiments for the removal of stumps have been going on for a long time in order to ascertain the commercial viability of the operation. Usually such experiments have concentrated on extracting the tree from the soil with roots intact (either by pulling it out of the ground or pushing it over) rather than felling the tree and then extracting the root system.

Tauber (1979) lists several advantages to such a system of logging operations. First, it would increase the output of timber from the main bole of the tree; this is because it would be possible to use the entire tree trunk down to the top of the root system, instead of having some of it attached to the root. Tauber estimates that usually up to 5 per cent of the total timber in the tree trunk is lost. Secondly, it would remove the necessity for stump removal in some circumstances as a sylvicultural measure prior to regeneration. If this were the case there would be additional economies possible in that mechanisation of soil preparation and seedling planting would be possible on a site already cleared of stumps. As well as that, stump removal would reduce the risk to the area from pathogens.

Although many machines have been experimented with in the past, no machine capable of root extraction as a part of the harvesting process has gone into production and found widespread use. Now a machine based on the LP-19 is being tested. This machine grips the trunk of the tree at its base and pulls the tree out of the ground bringing with it as it does so most of the root system. Then a system of knives severs the root ball from the trunk of the tree. It then forms bunches of felled trees in the same way as a feller-buncher but also creates small piles of roots. Although no figures are available, it is possible that the knife system for severing the root system is inferior to the chain saw cutting attachment used in the more usual feller-bunchers and so there might be some degradation of the butt-log, thus losing some of the advantage gained by harvesting a greater mass of cellulose.

Tauber (1979) estimates that, in the taiga of the North-west the root system comprises 27 per cent of the total biomass of spruce, 19 per cent for pine, 22 per cent for birch and 18 per cent for aspen. Although these appear to be substantial figures, they can only be gained by considerable effort. For example, in the Karelia area, the local research institute *KarNIILP* has shown that a tree with a diameter of 28 centimetres and a volume of 0.67 cubic metres would have a root volume of 0.13 cubic metres. Considering the energy required to retrieve this, it does not represent a particularly easily attained source of raw materials. Although it is not clear in the article whether this volume of root is the total volume or only the retrievable volume, it is true that much of the root system will remain in the ground with such a harvesting system. Roots less than 3 centimetres in diameter, which are likely to break off and stay in the ground, constitute up to a third of the total root system's mass. This is not altogether undesirable as it does mean that some part of the tree's organic material will remain in the soil and so will tend to reduce any possible soil degradation.

The major disadvantage to this method of increasing the output of the Soviet Union's forests is the fact that the root system contains a lot of dirt when it is extracted. Dirt, stones, etc. will blunt any mechanical instrument used to process the wood and will degrade any chemical agents used. Usually vibration techniques are used to get rid of any loose material and these can remove up to 70 per cent of the total and the rest is removed after the stumps and roots have been chipped by floating the chips off and allowing the debris to sink.

Tauber (1979) claims that an extra 35 to 49 million cubic metres of 'timber' are available if such a harvesting system became universal. Others disagree, however, putting the figure at 22 million (Sanikov *et al.*, 1980). Whatever the true figure it is unlikely that such a system of harvesting will find widespread use in the immediate future. The machines to fell trees with their roots are not in production and, if the long periods required for the machines discussed in Chapter 5 to get from the experimental stage to full production are indicative, they will not be in production for some time yet. The introduction of such machines would require the retiring from service of the new systems of feller-bunchers and feller-skidders only just going into service.

Crowns and Branchwood

Along with the consideration of stumps comes the idea of the use of crowns and small-sized branchwood in the timber processing industries. The use of such timber assortments presents less of a problem to the forest industries as crowns and branches are often transported to the lower landing if the full tree logging system is used. As this system is going to increase into the future so the amount of this type of timber available for conversion will increase.

The possibilities of converting such raw material into sawn timber is extremely small indeed, the limitations of size provide one constraint and also, as the timber in crowns and branches is physiologically younger, its characteristics are different to those of the timber contained in the main trunk of the tree and this also renders the timber unsuitable for such mechanical conversion. However, it is useful for other purposes.

The conversion of branchwood with a diameter greater than 3 centimetres has 'already been technically resolved' (Bikov, 1977), and may be converted into chips by the types of chippers now coming into use at lower landings. However, smaller branchwood than this remains practically unused although its volume is considerable. An extra 3.5 to 7 per cent of raw material is available if such branchwood were used. However, such small wood has the problem that it contains a relatively much larger proportion of bark and it comes in combination with a lot of leaf material.

According to State standards, the content of bark in the lowest grades of fibreboard should not exceed 15 per cent and for particle board the figure is 12 per cent and leaf material must be totally excluded. However, in the raw state, the small branchwood contains up to 30 per cent bark and over 50 per cent green material. Also the extraction system used is often such that dirt, mud and logging waste is caught up in the crowns of the trees as they are skidded across the forest floor. Often, the soil content of branchwood and crown systems after being skidded is in excess of 20 per cent whilst the State standard for usable crown wood is only 1 per cent. Such a situation is the worst of all states of affairs for the use of the present delimbing and chipping machines which have blades that will be rapidly blunted by the inclusion of dirt and debris and the reliability of which will be reduced below the present low levels, will be unsuitable without the installation of expensive cleaning apparatus.

At the moment there are no means for the debarking of small-sized branchwood and the techniques that are available for removing the bark after chipping do not, as yet, perform satisfactorily. So unless the norms for bark content in particle boards are relaxed there seems to be little use that can be made of this raw material at present.

THE EXPORT OF TIMBER AND TIMBER PRODUCTS

Introduction

With the increasing shortage of timber in the domestic economy during the coming years it is likely that the position of the Soviet Union as a major exporter of timber and timber products will come under increasing pressure. The opportunity cost of exports will increase as the domestic supply becomes ever more constrained and there will be pressure to divert supplies from international trade to the internal market.

The pattern of Soviet timber exports are shown in Table 9.8. To put these figures into some sort of global context, as far as sawn timber is concerned, the Soviet Union exports about half the amount that Canada does, but more than the USA, Sweden or Finland. This puts it into second place as an exporter of sawn timber in terms of volume if not value. However, its sales of processed timber products such as plywood, particle board and fibreboard are not so outstanding and its exports of these products are on a par with other smaller countries.

TABLE 9.8 Soviet Timber Exports

	1950	1960	1965	1970	1975	1976	1977	1979
Round timber[1]	0.7	4.4	11.0	15.3	16.9	17.9	17.4	15.2
Sawn timber[1]	1.0	5.0	8.0	8.0	7.8	8.6	8.1	7.6
Plywood[2]	48.3	129.1	190.0	281.0	302.0	319.2	337.0	314.6
Particle board[2]	-	-	60.0	145.0	273.0	271.1	320.1	227.0
Fibreboard[3]	-	-	12.0	41.0	72.0	81.1	88.8	98.6

1 - figures in millions of cubic metres
2 - figures in thousands of cubic metres
3 - figures in millions of square metres

Compiled from various sources

The distribution of the production of timber products for export is shown in Table 9.9. As would be expected, production is concentrated in the north west and the Far East; the bulk of the round timber exports originates from the Far East. The reason for this is the large expansion of the Japanese market for round timber. In the mid-1970s Japan took over 8 million cubic metres of round wood from the Soviet Union, a little under one-half of the total volume exported. Over the 1960s and 1970s trade with Japan showed a rapid expansion, of as much as 12 per cent per annum. The pattern of the trade with Japan, one of supplying a large amount of unprocessed timber, is to a large extent determined by the present lack of pro-

TABLE 9.9 Regional Distribution of Production for Export

	Percentage Share in Total Exports				
	sawn timber	round timber	plywood	particle board	fibre-board
North-west region	55.9	29.5	21.1	8.8	37.2
Central region	2.5	2.8	24.0	9.2	7.3
Volga-Vyatka region	6.5	6.6	6.3	9.3	8.2
Povolzhski region	0.0	0.0	8.5	0.0	0.0
The Urals	9.4	3.6	14.5	7.7	12.7
Western Siberia	0.8	0.9	6.3	8.8	6.9
Eastern Siberia	16.8	10.0	0.0	0.0	5.6
The Far East	3.3	45.6	0.0	0.0	0.0
Ukraine	0.8	0.2	3.2	27.8	0.0
White Russia	0.6	0.8	6.3	7.8	6.2
Latvia	1.4	0.0	6.7	1.4	2.2
Lithuania	0.0	0.0	0.0	9.5	13.7

Source: Kanevski and Shaitanov, 1975

cessing capacity in the Far East of the country. Indeed, the possibility of exporting timber from the Far East to Japan has been a stimulus in the expansion of the Far East's timber resources.

Trade in round wood to other countries has remained more-or-less constant in volume terms with Finland taking a large share of the remaining exports of this type and trade with Comecon also being significant.

Most of the round wood exported is in the form of saw logs and pulpwood (usually about 60 per cent and 25 per cent of the total round wood export respectively). The rest consists of pit props, building material and sleeper timber. Some veneer logs and round wood for match manufacture are exported, although these together account for only about 1 per cent of total round wood exports.

The production and export of sawn timber shows a different pattern; most of it is exported from the North-west region. Within this area about half of the sawn timber of *Archangelsklesprom* and *Karellesprom* is exported. Most of this goes by sea, especially through Archangelsk which is the Soviet Union's largest timber handling port. Rail transport is also used, especially from Karelia where about one-third of the timber is exported this way, and this means of transport is used to export to continental Europe and especially to the other eastern European countries.

Panel product exports from the Soviet Union are directed almost entirely to Europe and only a little goes to North America. The increase in exports of particle board from the Ukraine (see the table) is a comparatively recent phenomenon. The table also shows the absence of any significant exports of panel products from the Far East and East Siberia. This is again a reflection of the low level of industrial development of the region.

Trends and Prospects

Referring to Table 9.8, certain trends in the exports of forest products can be perceived. With the exception of fibreboard, the figures for 1979 are down on previous years. The large magnitude of the downturn is probably much the result of severe weather conditions in the winters of the late 1970s. However, apart from this, it is evident that, relatively speaking, panel product exports are expanding and round timber and sawn wood exports peaked in the early 1970s and are now stabilised or even declining.

There are good reasons for this trend, reasons which suggest that the pattern will continue. There is increasing pressure within the economy for timber and exports conflict with this. In the short-term the export contracts entered into by the Soviet Union take precedence. In the years 1977 and 1978, when the winters were especially severe (see Chapter 3) and transport of timber was held up, building work in the Archangelsk area was stopped as timber was diverted for export. Clearly this is not a long-term position.

In the long term, there is likely to be little potential for the future expansion of exports and, for reasons that have been discussed at length, the probability of a reduction. Given the need of the Soviet Union for foreign exchange to finance its imports of everything from grain to high technology, the ability to contract exports in any radical way is constrained. It is this pair of constraints between which the Soviet Union finds itself trapped. This gives rise to a strong incentive to substitute more processed products for round wood and sawn timber in the export mix so as to increase the value of the exports and it is this that leads to the pattern of exports shown in Table 9.8.

However, Kanevski and Shaitanov (1975) point out that this strategy is itself insufficient to guarantee increased export earnings coupled with a decreased usage of the timber resource. The effectiveness of substitution of sawn and round timber by panel products will depend on relative prices. Better, Kanevski and Shaitanov suggest, would be the export of higher value faced panel products, the production of which is still limited. However, as North and Solecki (1977) state, the sale of processed panel products '... is unlikely to be worthwhile as long as the buying countries have as much under-used processing capacity as they do now'. So the best strategy for the Soviet forest sector is likely to be constrained for the short to medium term, by external factors.

But such analysis suggests that the export of round timber to Western Europe is going to decline in the future as greater processing capacity comes into use. However, given the need for foreign currency, there is little chance for an internal expansion of the forest industries at the expense of exports.

10
The Wood Processing Industries

INTRODUCTION

In general, the structure of timber usage in the Soviet Union is more 'backward' than it is in the West in the sense that it exhibits an '... insufficient development in the production and demand for such ... material as paper, cardboard, panel products, plywood etc.' (Lobovikov and Petrov, 1976). Table 10.1 shows the production of various products from 1,000 cubic metres of timber harvested in different countries and illustrates this proposition.

TABLE 10.1 Production from 1,000 Cubic Metres of Harvested Timber

	Plywood M^3	Paper and Cardboard Tonnes	Particle Board M^3	Fibre-board M^2
Soviet Union	5.3	17.4	5.2	541
USA	42.2	134.3	12.1	2,124
Finland	15.7	95.0	8.4	1,540
Japan	118.5	260.0	7.1	1,764

Source: Varaksin and Stupnev, 1974

This 'backwardness' is also shown by the fact that in the Soviet Union about 15 per cent of the world's timber is cut, but only about 6.5 per cent of the world's plywood, 11 per cent of the world's particle board, 9.5 per cent of the fibreboard and 5 per cent of the newsprint, paper and cardboard products are produced. The difference in timber usage in the Soviet Union is most dramatically shown in the case of paper. Paper consumption in the Soviet Union is about 32 tonnes per 1,000 of population compared to about 80 in Western Europe and more than 230 in the USA (Lobovikov and Petrov, 1976). Whilst it is certain that paper production will increase in the future, it is never likely to reach the per capita levels of the West. There is always going to be less demand for wrapping and packaging and *Pravda* and *Izvestia* are never likely to become as large as Western newspapers.

To fulfil the aims of the eleventh five-year plan, 1981 to 1985, requires an increase in the output of particle board of 8.5 per cent per annum, fibreboard by 4.5 per cent per annum and paper by about 20 to 25 per cent over the full five-year period (Kirin, 1981). If there is no increase in the output of timber from the forests of the country, the raw materials needed will have to come from a substitution of timber from one use to another or by an increase in the amount of wood that can be used per cubic metre of timber harvested. The plan would require an increase in output from the country's forests, if it were to be fulfilled entirely by the use of timber not reallocated from other uses or by an increase in the overall conversion ratio, of about an extra 1 per cent per annum.

The uses of the timber harvested in the Soviet Union are shown in Table 10.2. These figures represent the volume of usable or commercial timber used for various purposes. Thus, in 1960 when 369.5 million cubic metres were harvested, 263 million were used for commercial purposes - the remaining 106.5 million, or 28.8 per cent, went as waste or firewood. By 1975 the proportion going for these uses had fallen to 17.7 per cent and the volume had fallen to only 70 million cubic metres. The process of changing from using timber for industrial purposes rather than fuel offers a means of increasing output of forest products without, at the same time, increasing the volume of timber removals. This was discussed in Chapter 9 where it was seen that important constraints were developing that would make this strategy difficult to pursue with any vigour.

TABLE 10.2 Uses of Timber

	1960		1975	
	%	mill. M^3	%	mill. M^3
Sawn timber	58.0	151.7	58.9	191.7
Plywood	2.6	6.8	3.4	11.0
Particle board	neg	0.6	4.5	14.6
Fibreboard	0.1	1.4	2.7	8.7
Paper and cellulose	5.0	13.0	7.9	25.6
Mining needs and charcoal production	6.5	17.0	4.0	13.0
Transmission poles	1.9	4.9	1.2	3.9
Boxing and pallets	1.9	4.9	1.8	5.8
Use within the logging industry	6.5	17.0	4.1	13.3
Matches	0.2	0.5	0.3	1.0
Other	17.3	45.2	11.2	36.4
TOTAL	100.0	263.0	100.0	325.0

In Table 10.2 it is evident that the proportion of timber used for sawn timber has remained about constant. Increases have been seen in the use of timber for the production of plywood and paper, but most significantly for the manufacture of particle board and fibreboard. The figures in this table show a decline in some uses of timber. Less timber is used for the mining industry in the form of pit props than has been used in past years. It is expected that this trend will continue. The use of wood for transmission poles has been declining. Significantly, the use of timber within the forest industries has declined, partly for reasons that have been discussed at length in this book.

The uses to which the main products of the wood processing industries are put are shown in Table 10.3. A significant proportion of the sawn timber is used in construction. The use of panel products in construction is less developed in the Soviet Union than it is in the West and here savings in timber could be made by the use of panel products where possible rather than sawn timber. This linkage between timber production and construction is one which connects the forest industry to all areas of the economy. Fixed investment in almost anything will require a timber input and so a constraint in timber output will slow down economic growth and will have effects on industry and housing programmes.

TABLE 10.3 Uses of the Major Wood Products

	Percentage Use of			
	sawn wood	plywood	particle board	fibre-board
Construction	36.7	4.1	12.5	24.0
Repair and maintenance work	12.2	1.5	-	2.0
Furniture	5.0	36.5	75.0	25.0
Engineering	4.4	8.0	-	7.0
Packaging	9.1	11.0	-	7.0
Charcoal	1.4	-	-	-
Others	31.2	38.9	12.5	25.0

Source: Glotov, 1977

SAWMILLING

Sawmilling is a very important user of timber, taking nearly two-thirds of the output of useful timber from the country's logging concerns. The output of sawn timber has remained more-or-less constant over recent years at just over 100 million cubic metres of sawn timber per annum, a level which it first reached in 1960. The problems faced by the sawmilling industry are similar in their basic nature to those seen in the logging industry. The production is not concentrated into large efficient units and the location of those units is not satisfactory.

Concentration

As for the first point, of the 116.2 million cubic metres of sawn timber produced in the Soviet Union in 1975, only about 44 million were produced in concerns coming under the control of the main ministry. The remainder was produced in enterprises coming under the control of forty-six other ministries and departments (Stupnev *et al.*, 1976). This position may have been improved with the recent merger of the old Ministry for Timber Industries and the Ministry for Pulp and Paper Production into one larger unit - *Minlesbumprom*. But the position is clearly unsatisfactory from the point of view of rational planning and complete use of the raw material. If more than half of the country's sawn timber is being produced in enterprises not under the direct control of the Ministry the planning for all the enterprises producing sawn timber will only be considered as a whole at *Gosplan* level; this will mean that production cannot really be concentrated to exploit economies of scale and the use of wastes from sawing for the production of other products is going to be inhibited if the wastes occur in enterprises under the control of one ministry while they could be used in the concerns of *Minlesbumprom*.

The problem of concentrating production is shown by one or two figures. There are about 32,500 sawmills in the Soviet Union. This gives an average production of 4,960 cubic metres per annum. This compares with about 11,000 in the USA and 27,000 in Canada, although it is not small compared to some European countries. The distribution of sizes of sawmilling enterprises shows a very skewed pattern. There are very many very small sawmills and a few large units. About 350 units in the country have annual outputs of more than 50,000 cubic metres and most of these are under the control of *Minlesbumprom*.

Most production indicators of the sawmills of the enterprises not under the control of the Ministry are worse than those of the plants that are. One cubic metre of sawn timber costs 35 roubles to produce in the sawmills of *Minlesbumprom* compared to 50 to 55 roubles for the other enterprises. *Minlesbumprom's* sawmills work to 95 per cent of full capacity, while the others operate at only 60 per cent (Kudriavtsev, 1975). Also, overall, the labour input into sawn timber production in the Soviet Union is considerably higher than it is in other countries. In the early 1970s, according to one Soviet author, the figures were as follows - in the Soviet Union 1 cubic metre of sawn timber required 6.08 man hours compared to 2.36 in Canada and 2.56 in Sweden (Kudriavtsev, 1975).

The Location of Sawmilling

The geographical pattern shown by the production of sawn timber is not unlike that of logging activity. Figures for the distribution of sawn timber production are given in Table 10.4. According to these figures there has been a gradual change in the location location of sawmilling from the lightly forested to the heavily forested area. Clearly, all other things being equal, it would

TABLE 10.4 The Location of Sawn Timber Production

	1940		1950		1960		1975	
	mill. M^3	%	mill. M^3	%	mill. M^3	%	mill. M^3	%
Soviet Union	34.8		49.5		105.6		116.2	
Heavily forested area	17.1	49.1	21.4	43.2	50.9	48.2	64.2	55.2
European part	6.5	18.7	7.7	15.6	19.1	18.1	21.6	18.6
Urals	3.3	9.5	5.2	10.5	10.7	10.1	11.5	9.9
Western Siberia	2.1	6.0	2.3	4.6	4.4	4.2	7.1	6.0
Eastern Siberia	2.8	8.0	4.1	8.3	11.6	11.0	17.4	15.0
The Far East	2.4	6.9	2.1	4.2	5.1	4.8	6.6	5.7
Lightly forested areas	17.7	50.9	28.1	56.8	54.7	51.8	52.0	44.8
North west	1.5	4.3	2.3	4.6	3.6	3.4	3.3	2.8
The Baltic States	0.6	1.7	2.1	4.2	3.0	2.8	3.2	2.8
White Russia	1.6	4.6	2.3	4.6	3.0	2.8	3.2	2.8
The central area	5.8	16.7	8.0	16.2	15.2	14.4	13.3	11.5
The Povolzhski Region	2.2	6.3	2.5	5.1	6.2	5.9	5.4	4.6
The Ukraine	3.0	8.6	6.2	12.5	11.0	10.4	10.4	8.6
The Caucasus	1.0	2.9	1.9	3.9	4.4	4.2	4.1	3.5
Urals and Western Siberia	1.6	4.6	1.9	3.9	4.9	4.5	5.5	4.7
Soviet Central Asia	0.4	1.2	0.9	1.8	3.4	3.2	4.0	3.5

Source: Glotov, 1977

be better to saw timber at the point where it is felled rather than in the area it is to be used as the transport costs of 1 cubic metre of sawn timber will be less than the costs of transporting enough timber to produce 1 cubic metre of sawn timber.

In 1940 a little over half of the country's sawn timber was produced in the lightly forested areas with the most significant production being in the Ukraine (8.6 per cent of the total) and the Central Region (16.7 per cent). By 1975 the lightly forested area's output of sawn timber had increased by more than 34 million cubic metres but its share in the country's total output had fallen to only 44.8 per cent. This relocation of production has not been a smooth process. In 1950 the importance of the lightly forested area increased compared to 1940 - with a notable increase in output in the Ukraine, the Central Region and the Baltic States. The reason for this was that the destruction of industries and the housing stock during the war meant that rebuilding was of the utmost urgency and this meant timber was required for the necessary construction. The easiest and quickest way to accomplish this was to build sawmilling capacity in the regions where the timber was needed most - the Ukraine, the Baltic States and Central Region. This increase in the output of sawn timber from these areas was the result of the rise of many small scale single frame sawmills. These types of mills were used not only in the European part of the Soviet Union but in the areas in the heavily forested regions too. The 1950s and early 1960s were the times of heavy over-cutting in the forests in the European parts of the country. The allowable cut, especially for coniferous species, was exceeded by very large margins in this area - see Table 3.5.

Serious investment in sawmilling capacity in the heavily forested areas began in the mid-1950s with the sixth five-year plan. It was at this time that the large sawmilling capacity in the Bratsk, Krasnoyarsk areas and around the Yenisei in the east as well as in the north of the European part of the country around Archangelsk was built.

But, despite the fact that the investment in sawmilling capacity began so long ago, there are still capacity constraints in the production of sawn timber in the areas where logging is being developed. In Chapter 2 it was seen that in the North-West Region, Western Siberia, Eastern Siberia and the Far East all export timber to other areas to service their processing capacity. For example, the Eastern Siberia region exported in 1975 about 11 million cubic metres of round wood and only 6.5 million cubic metres of sawn timber. Some of the round wood was exported all the way to the Ukraine. In total, round wood for sawmilling is transported on average some 2,000 kilometres, giving a total transport requirement of 360,000 million cubic metre kilometres (Varaksin and Stupnev, 1974).

Some Technological Considerations

The short-term considerations of the immediate post-war five-year plans have been mentioned; the policy of these plans to rebuild rapidly the productive capacity in the west of the country led, in a sense, to the present problems faced by the industry of production being scattered around many small mills. From the eighth five-year plan, beginning in 1966, the policy began to change. From that date a stance was taken to increase the total utilisation of the wood delivered to sawmills and to modernise the industry.

Beginning at this time there was the installation of some debarkers in sawmills. This has increased, but even now only a little less than 20 million cubic metres of the timber received by sawmills (about 10 per cent of the total) is debarked; most of this goes for export. Experience shows that the productivity per shift of saw lines is increased by about 5 to 8 per cent if debarked timber is used. Also, the cutting edges of blades are blunted more slowly because of the fact that they do not have to cut through the bark which often contains dirt from logging. The quality of timber produced by sawing debarked timber is usually considered to be higher and so it is concluded that debarking is suitable for '... all raw material under all circumstances' (Varaksin and Stupnev, 1974). Most timber is debarked by drum debarkers of domestic production, but imported models are in limited use.

Preservation capacity was also installed, but only on a modest scale. It is calculated (the methodology is not presented and so the figure must be taken at face value) that preservation saves about 75 kopecks per cubic metre of sawn timber treated (Varaksin and Stupnev, 1974). Presumably these savings come in the form of longer life of the timber used and so a reduction in replacement and repair. But despite this saving, only about 3 to 4 million cubic metres of sawn timber are treated each year in the sawmills of *Minlesbumprom* - although this does represent about 9 per cent of the Ministry's output. It is likely that the other sawmills, under the control of the other ministries, preserve less than does *Minlesbumprom*. In the years up to the mid-1970s, 67 enterprises of the Ministry had kiln seasoning installed with a capacity of 2.2 million cubic metres of sawn timber per annum. In all, about 40 per cent of the sawn timber from *Minlesbumprom* is seasoned in some way or other; about 15 to 20 per cent is kiln dried and the rest is seasoned atmospherically. Some complain of the old style construction of the kilns that many sawmills have been supplied with, in some it is almost impossible to regulate the drying process and the operations of loading and emptying of the kilns has to be carried out by hand. By 1985 it is hoped that over half of the sawn timber produced will be kiln dried and a quarter seasoned atmospherically.

However, all of this investment did not significantly affect such indicators as the productivity of labour to any marked degree. Part of the reason for this was the lack of volume of investment. By 1971 the capital intensity of the main Ministry's sawmilling industry was 3.8 thousand roubles per worker compared to 7.2 for the industrial average for the country as a whole (Varaksin and Stupnev, 1974). Whilst indicators like this can be misleading, it is probably correct to interpret this disparity (nearly two to one) as expressing a capital deficiency in sawmilling.

So there seems to be a lot that can be done to increase the seasoning and preservation of timber. But the most pressing aims of the industry, judging by the objectives of the most recent plans, are to increase the total usage of the raw material supplied and also to raise the productivity of labour. The reasons for these aims are very much the same as the reasons behind similar objectives

TABLE 10.5 The Use of Material Received at Sawmills (millions of cubic metres)

	1965	1970	1975
Raw material received	62.0	69.0	72.0
Output - total	43.4	51.8	56.5
of which: sawn timber	40.1	43.8	44.9
chips for paper making	0.8	3.2	4.3
chips for panel products	2.5	4.8	7.3

Source: Glotov *et al.*, 1975

in the logging industry, that is a labour shortage and a need to increase the efficiency of production and concentrate on intensive expansion as opposed to extensive.

The conversion ratios of the raw material supplied to sawmills has been increasing as can be seen in Table 10.5. These figures refer to the output of *Minlesbumprom* only. According to the data, the overall conversion ratio of raw material input into output of some form has increased from 70 per cent in 1965 to 78.5 per cent in 1975. This has come about, however, by the use of waste for production of chips, the output of which has increased by 8 million cubic metres. The conversion ratio of input into sawn timber has, in fact, fallen from 64.7 per cent in 1965 to 62.4 per cent in 1975. This decline means that even a maintenance of sawn timber output at a given level will require the delivery of an ever greater amount of round wood. The reason for this slow decline in the efficiency of sawmilling in terms of the production of sawn timber lies in the deterioration of the raw material supplied. As average tree sizes decline because less and less favourable areas are logged over, the conversion ratio will decline.

If broadleaved species are to be more fully utilised it is likely that conversion ratios will fall sharply as their inherently poorer stem form will mean that more waste will be generated. At the present about 85 per cent of the sawn timber is coniferous. Most of the remaining 15 per cent is made up of birch and aspen and only a small proportion (about 2 to 3 per cent) is made up of oak and beech. It seems that the broadleaved sawn timber produced is of a lower quality than average if its end use is indicative. Only 11.5 per cent of it is used for construction, compared to 36.7 per cent over all - see Table 10.3. About 24 per cent of the broadleaved sawn timber is used for packaging and pallet construction, compared to only 9.1 per cent overall. However, it is in furniture construction that the better quality of some broadleaved sawn timber is shown, and in this area much more sawn timber of broadleaved species is used compared to coniferous.

The figures in Table 10.5 show that in 1975 27.1 million cubic metres of waste were generated in the sawmills of *Minlesbumprom*, of which some was used for the manufacture of chips. In the industry as a whole the figure is commensurately greater. Of the 27.1 million cubic metres of waste in the main Ministry's operations, about 17 million are in the form of edges, off-cuts, narrow laths, etc., and most of the remainder is sawdust. However, the use of these wastes is limited by the organisation of the sawmilling industry. Table 10.5 shows that, in 1975, in the sawmills of the main Ministry, 4.3 million cubic metres of chips were produced for the production of paper. Over the five years 1971 to 1975, the total volume of chips produced for this purpose was a little over 18 million - about twice the volume produced in the preceding five-year period. But during the ten-year period the sawmilling concerns in all of the other ministries and departments converted only about half a million cubic metres of waste into chips (Stupnev *et al.*, 1976).

Of the sawdust produced only about 1 million cubic metres was used in the wood chemical industry and only a small quantity of the rest was used for industrial purposes; most of the remainder was burnt. This was the fate of most of the bark produced in debarking operations.

As far as technological plans are concerned, the disadvantages of frame saws are recognised. These provide the mainstay of sawing capacity in the country but they are a slow means of sawing timber. Plans for new investment involve the use of circular saws or band saws which offer higher cutting speeds and hence higher productivity. With the installation of capacity of produce particle board, profile chippers are to be used. The average throughput of the mills is to be increased with the introduction of the improved saws.

PLYWOOD MANUFACTURE

The manufacture of plywood has a long history, and its production in Russia began at the end of the nineteenth century. Although many species are used throughout the world for plywood manufacture, in the Soviet Union birch is usually used, making up 85 per cent of the plywood output. As well as birch, oak, alder and lime are used, and from the coniferous species, pine and some larch are occasionally made into plywood.

Locational Factors

The location of plywood manufacture has not really altered radically since 1960. In Table 10.6 it is seen that in 1960 61 per cent of all plywood was produced in the lightly forested areas and in 1975 the figure had fallen to 58 per cent. Investment has been carried out in that time as output overall has increased, but it has not been concentrated in the heavily forested regions.

The result of this is the familiar story that in the European lightly forested parts of the country, the timber supply is limited and future expansion of output is difficult to visualise. In Chapter 3 it was seen that the broadleaved allowable cut was often under-utilised, even in European areas, so the raw material position might not be thought to be as pressing as it is for, say, the production of sawn timber. However, the unused broadleaved allowable cut may represent timber unsuitable for plywood production, for example, it might be too small. Glotov (1977) states;

TABLE 10.6 Location of Plywood Manufacture

	1940 thousand M^3	%	1960 thousand M^3	%	1975 thousand M^3	%
Soviet Union	731		1,354		2,196	
Heavily forested area	152	21	527	39	915	42
inc. North Europe	108	15	339	25	430	20
Urals	0	0	108	8	335	16
Western Siberia	23	3	32	2	54	2
Eastern Siberia	7	1	17	1	50	2
Far East	14	2	31	2	46	2
Lightly forested area	579	79	827	61	1,281	58

Source: Glotov, 1977

> 'At the present moment, free raw material for the production of plywood in the European part of the country is limited and the possibility of further expansion (of plywood manufacture) has been exhausted. Also, at the present moment, many plywood plants in the lightly forested regions use timber brought in from great distances.'

Indeed, the worsening of the nature of the raw material, most specifically in terms of the average size of the trees supplied, is blamed for a slight reduction in the efficiency of plywood manufacture recently. In the late 1960s about 2.5 cubic metres of timber was needed for 1 cubic metre of plywood, but now this figure has increased to 2.7, with a range of between 2.3 to 3.

Technological Points

Most of the new investment in panel products in the future is to go into the production of particle board and fibreboard, with no radical increase in the output of plywood foreseen.

However, some changes are planned. To begin with, it is expected that the sizes of plywood sheets produced are to become more varied. At the present about 95 per cent of plywood is produced in one standard size - 1525 millimetres square. Also, it is hoped to increase the amount of exterior grade plywood to between one-third and one-half of the total output. At present little of the output is of this type which inhibits the use of plywood in many exterior applications.

It is also hoped that investment in plant will lead to a reduction in the amount of timber used to produce plywood. It has already been stated that this figure has increased on average to about 2.7 cubic metres per cubic metre. A figure of between 2 to 2.3 is being aimed at. Technological progress is expected to reduce the labour input as well. At the present, a little more than

34 man-hours per cubic metre of plywood are required, but this figure varies widely, from 13 to 40. The main problem, as with other branches of the timber industries, is the large amount of manual labour and physical handling of the raw material and final product.

PARTICLE BOARD AND FIBREBOARD

These two products represent a relatively new use of timber in the Soviet Union. This is evident in the fact that production of these boards in any quantity only started in the 1950s. They represent a branch of the forest economy that is bound to expand quite vigorously into the future. They have the advantage over other uses of timber in that the raw material used for their production can be of low quality. This means that the decline in the quality of the growing stock is not something which will affect the production of these boards as it will the production of plywood and sawn timber. Also the possibility of using waste from the manufacture of sawn timber, plywood and wood otherwise left unconverted exists.

There are over 85 mills in the Soviet Union engaged in the manufacture of particle board; most are under the control of *Minlesbumprom* and these account for about 85 to 90 per cent of the total output. This is not surprising given the greater capital intensity attached to the production of particle board. Also, some of the raw material for the board comes from waste from sawmills - a raw material which the Ministry should have in abundance. About 8 to 10 per cent of the raw material used in particle board is in the form of waste from sawmilling. The main sources of wood however, between 65 and 67 per cent, comes from the use of low quality wood and 'firewood quality' timber. A significant percentage comes from the use of waste from the veneering and plywood industry. This supplies almost 10 per cent of the raw material input.

This ability to make use of low quality wood and the waste from other forest industries means that there is less of an incentive to locate these industries in the Far East and Siberia. It is possible to locate particle board and fibreboard mills close to existing sawmills and not generate any greater basic timber demand. The location of board manufacture is shown in Table 10.7. These figures naturally show a rapid increase in the production of these boards between the years 1960 and 1975. As well as that there has been a change in the weight of production in the fifteen years. In 1960 about 94 per cent of particle board production was situated in the lightly forested area, whilst in 1975 the percentage had fallen to 72. This was the result of a growth in production of about 38 per cent per annum in the heavily forested parts of the country compared to only 22 per cent in the lightly forested areas. For fibreboard the figures for annual increases in production are 13.8 and 11.4 per cent between 1960 and 1975 in the heavily and lightly forested areas respectively. However, a lot of the development in processing capacity has occurred in the European parts of the heavily forested area - that is around Archangelsk and in the Komi ASSR. Development in the Far East and Siberia is still low, only 8.5 per cent of particle board is produced there.

There is still capacity for expansion in the European part of the country, which is desirable as the main areas of demand are

TABLE 10.7 Location of Particle Board and Fibreboard Manufacture

	Particle Board[1]		Fibreboard[2]	
	1960	1975	1960	1975
Heavily forested area	9	1,108	33	230
European part	6	468	17	101
Urals	1	303	11	55
Western Siberia	2	97	0	8
Eastern Siberia	0	174	0	48
The Far East	0	66	5	18
Lightly forested area	152	2,886	35	178
North west	40	203	10	17
The Baltic States	14	225	0	30
White Russia	7	256	2	29
The central area	31	751	20	55
The Povolzhski Region	1	212	0	22
The Ukraine	48	806	3	25
The Caucasus	11	361	0	0
Soviet Central Asia	0	72	0	0

Source: Glotov, 1977

1 - thousands of cubic metres
2 - millions of square metres

located there. The major consumer is the furniture industry. This takes almost three-quarters of the particle board output with construction work taking about 17 per cent. This is a relatively low proportion; in Western countries the figure is much higher. There are economies to be had in furniture production in the use of particle board. In the furniture industry, according to Soviet figures, 1 cubic metre of particle board can replace 2.4 cubic metres of sawn timber and whilst the particle board would only require the conversion of 1.5 to 1.8 cubic metres of low quality wood, the sawn timber needed would take 3.7 cubic metres of saw logs to produce.

As with plywood, the variety of particle board and fibreboard produced in the Soviet Union is limited and it is intended to expand the range in the future. In the past some extruded particle board was manufactured. In 1965 it made up 4.1 per cent of the total output and in 1975, 3 per cent. But the low strength properties of such a board necessitates its covering with a veneer and this is not always economical. No more investment in the production of this type of board is planned (Varaksin and Stupnev, 1974).

Glossary

Allowable cut - the amount, determined by *Gosleskhoz*, that can be removed from the forests of the country or a specific area such that a sustained yield is possible.

Bole - the main trunk of the tree.

Broadleaved species - in this book this term is used to cover all non-coniferous species, either the tree or the timber from it, e.g. birch, oak, aspen, etc. In Soviet usage such species are divided into *tvyordolistvenni* (hard wood non-coniferous species, e.g. oak, beech) and *myakolistvenni* (soft wood non-coniferous species, e.g. birch, aspen).

Coniferous species - for example pines, spruces, larches, firs, etc. Note that this term is not the 'opposite' of deciduous trees as the larch, which bears cones, is deciduous.

Crown - the leafy upper part of the tree.

Current annual increment - the change in a parameter describing the trees in a given area of forest during a specific year, usually it refers to the increase in volume per hectare in a given year.

Discount rate - the percentage rate at which the present value of a future income stream declines. Thus, if the discount rate is 10 per cent, the value now of 110 roubles in one year's time is 100 roubles.

Extraction - the second stage of timber transportation, occurring after skidding. The process of moving timber from the upper to the lower landing, or sometimes, to the final user.

Feller-buncher - a machine designed to fell trees and form the felled trees into loose unbound bunches of several trees which are then left on the forest floor for skidding by a separate tractor.

Feller-skidder - a machine designed to fell trees and skid them over the forest floor. Usually skidding is carried out after several trees have been felled.

Fibreboard - there are two types of fibreboard. They are formed by the conversion of wood into coarse pulp consisting of bundles of particles rather than individual fibres as would be the case with paper. The pulp is then formed into a thick mat which is dried at high temperature. The drying is carried out either with high pressure on the mat to form hardboard or without pressure when the product is the lighter, less dense, insulation board.

Five-year plan - the basic nation-wide plan which is published every five years (the eleventh f.y.p began in January 1981) and is designed to show the development of all of the economy over the period in question.

Full tree system - sometimes known as the whole tree system, this is a system of logging which involves the extraction to the lower landing of trees which have not been delimbed. The only operations carried out on the forest floor and at the upper landing are felling, skidding and loading.

Giprolestrans - a Leningrad-based research and testing agency looking mainly at problems of logging.

Growing stock - the volume of timber growing on a particular site.

Intermediate cutting - the removal of timber prior to the final felling of the mature crop.

LenNIILKHh - a Leningrad-based scientific institute for research into sylviculture. Its work concentrates on aspects of forestry other than logging of the major forests and looks at the technology and economics of the establishment of plantations, thinnings, site amelioration, etc.

Mean annual increment - a measure of a parameter describing the trees in a given area of forest expressed as an arithmetic annual average amount. Usually, this is the average annual volume increment per hectare. In this case it will equal the total volume of the growing stock on a hectare divided by the age of the forest.

Particle board - a panel product produced from chips of wood bonded together into a rigid material by the means of a resin binder. A faced board is a board that has a facing of veneer.

Plywood - a board produced by the gluing of veneers of timber together with grain orientation in alternate directions.

Price c.i.f. - the 'price including carriage, insurance and freight'. Here, the price that a timber user must pay for 1 cubic metre of timber to be delivered to him.

Price f.o.b. - the 'price free on board'. Here, the price at the *lespromkhoz* that is received by the logging concern for a unit of timber.

Rotation length - the length of time between the planting or regeneration by some means of a forest and the final felling of the trees.

Short wood system - a system of logging which usually involves the delimbing and cross-cutting of a tree on the forest floor or at the upper landing. Short log lengths are then extracted.

Skidder - a tractor designed for skidding timber.

Skidding - the initial stage of timber transportation, moving timber from the site where it is felled to the upper landing.

Sylviculture - the study of trees, their growth, form and the effects of different operations on trees when grown in forests or woods.

Techpromfinplan - the one-year plan that an enterprise receives detailing the operations of that enterprise for the year in question.

Thinnings - operations to remove timber, not necessarily for sale, prior to the final felling of the mature crop.

Tree length system - a system of logging which involves the delimbing of a tree on the forest floor or at the upper landing. Tree length logs are then extracted to the lower landing.

TsNIILesosplav - The Central Scientific Institution for Research into Timber Floating and the Water Transport of Timber.

TsNIIME - the Moscow-based Central Scientific Institution for Research into the Mechanisation and Energy Needs of the Forest Industries. Its research tends to concentrate on the technology and, to a lesser extent, the economics of the logging process and work at the lower landing.

Under-storey - a crop of young trees growing below the main crop of mature timber. Only certain species of tree can grow in their own shade and so an understorey does not always occur.

Upper landing - a site in the forest where timber is transferred from the skidder, which is slow and unsuitable for long hauls, to the means of transport which will extract the timber to the lower landing.

NB Many transliterated Russian terms are used in the text, e.g. *lespromkhoz*, and these are explained in the text. The terms applying to the organisation of logging are described in Chapter 3.

References

Abbreviations used:

LELS - Fsyesoyuzni Nauchno-issledovatelni i Proektni Institut Ekonomiki, Organizatsii Upravleniya Proizvodstvom i Informatsii po Lesnoi, Tselluloznobumaznoi i Derevoobrabativaushchi Promishlennosti, 'Lesoekspluatatsi i Lesosplav'.

A journal published thirty-six times a year in Moscow.

LKz - Lesnoe Khoziastvo.

A monthly journal published in Moscow. It deals mainly with sylvicultural matters and their related economics.

LP - Lesnaya Promishlennost

A monthly journal published in Moscow. It deals mainly with matters concerning the logging industry.

LPPH - Lesnaya Promishlennost Publishing House

A publishing house in Moscow specialising in forestry textbooks.

VNIPIEILesProm Ek Info

- Fsyesoyuzni Nauchno-issledovatelni i Proektni Institut Ekonomiki, Organizatsii Upravleniya Proizvodstvom i Informatsii po Lesnoi, Tselluloznobumaznoi i Derevoobrabativaushchi Promishlennosti, 'Ekspress Informatsia'.

An information sheet published in Moscow.

Abol, P.I., Zalkind, A.S., Kushankin, N.F., Levin, S.M. and Lumanov, R.A., (1980) Valochno-trelevochnaya Mashina LP-49 (The LP-49 Feller-skidder) LP, April, p.11.

Alekhin, L.V., (1976) LO-72 Ukrepenoi Brigadye (The LO-72 in an Enlarged Brigade) LP, January, p.7.

Alekseyev, L.A., (1982) Pyatilyetka Lesozatovitelei (Loggers' Five-Year Plan) LP, April, p.1.

Algvere, K.V., (1966) Forest Economy of the USSR, Studia Forestalia Suecica No. 39, Stockholm.

Andreyev, V.S., (1976) Sovietskie Trudovie Prava (Soviet Labour Law) Vishaya Shkola, Moscow.

Anon (1947) Lesnaya Promishlennost za 30 Let (The Forest Industries over Thirty Years) LP, October, p.5.

Anon (1949) Pervi Opit Treliovki Traktorami KT-12 (First Experience of Skidding with the KT-12) LP, January, p.11.

Anon (1960) Tipovia Metodika Opredelena Ekonomicheskoi Effektivnosti Kapitalnikh Vlozhenii i Novoi Tekhniki v Narodnom Khoziastve SSSR (The Recommended Method of Calculating the Economic Effectiveness of Investment in New Technology in the USSR) Gosplanizdat, Moscow.

Anon (1967) Preiskurant No 07-03; Optovie Tseni na Lesoproduktsi (Price List No 07-03; Timber Prices) Preiskurantiz, Moscow.

Anon (1968) Opredelenie Ekonomicheskoi Dostupnosti Resursov Nizkosortnoi Drevesini i Drevesnikh Otkhodov Lesozagotovok (The Definition of Economic Availability of Low Quality Wood and Logging Waste) Lesoteknicheskaya Academia Nauchni Trudi No 116, Leningrad.

Anon (1969) Tipovia Metodika Opredelenie Ekonomicheskoi Effectivnosti Kapitalnikh Vlozhenii (The Recommended Method of Calculating the Economic Effectiveness of Investment) Ekonomika, Moscow.

Anon (1973) Katalog Sovietskikh Traktorov (Catalogue of Soviet Tractors) Gosudarstvenni Soyuzni Nauchno-issledovatielski Traktorni Institut, Moscow.

Anon (1974) Spravochnik Ekonomista Derevoobrabativauschi Promishlennosti (A Handbook for Economists in the Woodworking Industry) LPPH, Moscow.

Anon (1977a) Mashtabi Rosta (Rates of Growth) LP, August, p.6.

Anon (1977b) Mashtabi Rosta (Rates of Growth) LP, September, p.5.

Anon (1977c) Mashtabi Rosta (Rates of Growth) LP, October, p.12.

Anon (1977d) Opredelenie Effekivnosti Novoi Tekhniki (The Definition of the Economic Effectiveness of New Techniques) Ekonomicheskaya Gazeta, March.

Anon (1979) Na Novi Rubazhi (New Commitments) LP, April, p.1.

Anoshin, R.M., Voronin, I.V., Pavlov, V.V., Kulikov, T.A., and Tryanov, M.A., (1977) Ekonomika, Organizatsia i Planirovania Proizvodstva v Leskhozakh (The Economics, Organisation and Planning of Production in *Leskhozi*) LPPH, Moscow.

Anuchin, N.P., (1977a) Teoria i Praktika Operatsii Lesnova Khoziastva (The Theory and Practice of Forestry) LPPH, Moscow.

Anuchin, N.P., (1977b) Analiz Vozrasta Rubok i Raschot Lesopolzovania (Analysis of the Rotation Length and the Calculation of the Allowable Cut) LP, December, p.24.

Artamov, M.D., and Michaelovski, U.V., (1975) Traktori na Lesozagotovok (Tractors in Logging) LPPH, Moscow.

Asonov, A.A., (1979) Predpushkovaya Podgotovka Mashin Zimoi (Preparing Equipment for Winter) LP, October, p.14.

Bagayev, N.G., (1976) Vstrechaya Partini Syezd (Meeting the Party Congress) LP, February, p.1.

Bagayev, N.G., (1979) Effektivnost Shtabelovki Khlistov (The Efficiency of Storing Tree Lengths) LP, February, p.4.

Baranovski, V.A., Brik, M.I., and Burdin, N.A., (1976) Tekhnicheskoi Razvitie Lesnoi Promishlennosti (The Technical Growth of the Logging Industry) LPPH, Moscow.

Baranovski, V.A., (1977) Puti Snizhenie Trudoyomkosti Lesozagotovitelnikh Rabot (Ways to Reduce the Labour Content of Logging) LP, March, p.1.

Batrakov, V.P., (1980) Lesnim Stroikam - Chetki Ritm (Rhythmical Work on Logging Projects) LP, June, p.1.

Belenov, I.A., (1980) Lesosplav i Okhrana Vodoyomov (Timber Floating and the Protection of Reservoirs) LP, March, p.4.

Bikov, E.N., (1977) Promishlennoi Izpolzovanie Vetki i Suchev (The Industrial Use of Branch Wood) LP, April, p.14.

Blandon, P.R., (1978) von Thunen's Theory of Land Use and the Economics of Forestry, City of London Polytechnic Working Paper, London.

Boldanov, P.L., (1974) Dendrologia (Dendrology) LPPH, Moscow.

Borisoviets, V.P., (1977) Splavshchiki - Narodnomu Khoziastvu (Timber Floaters Need to Make a Contribution to the National Economy) LP, October, p.17.

Borisoviets, V.P., (1979) Navstrechii Splavnoi Navigatsii (The Beginning of the Timber Floating Season) LP, March, p.1.

Borisoviets, V.P., (1980) Lesosplav-80 Pered Startom (The 1980 Timber Floating Season, Before the Start) LP, March, p.1.

Borisoviets, V.P., (1982) Lesosplav-82: Tempi, Organizovannost, Kachestvo (1982 Log Floating Season: Speed, Organisation and Quality) LP, March, p.2.

Borski, N.E., (1976) Kompletki Mashin Rabotaut v Dvye Smeni (Groups of Machines in Two Shift Working) LP, February, p.6.

Bozhak, V.L., (1968) Valochno-paketirushchaya Mashina Dyatel-2 (The Dyatel-2 Feller-buncher) LP, May, p.5.

Bronina, A., and Chubais, B., (1969) Trudovie Prava Rabotnikov Lesnoi Promishlennosti (Workers' Rights in the Logging Industry) Profizdat, Moscow.

Burdin, N.A., (1973) Effektivnost Raboti Valochno-paketirushchi Mashin LP-2 (The Effectiveness of the LP-2) LELS, No.14, p.5.

Burdin, N.A., (1975) Effektivnost Primenenie Novoi Tekhniki na Lesozagotovok (The Effectiveness of the Use of New Technology in Logging) LP, November, p.27.

Bursin, E.E., (1977) Tsenoobrazovanea v Lesopilnoi i Derevoobrabativaushchi Promishlennosti (Price Formation in the Sawmilling and Woodworking Industries) LPPH, Moscow.

Bursin, E.E., (1978) Prodolzhit Sovershenstvovania Tsen na Lesoproduktsi (Continue the Improvement of Prices for Timber Products) LP, May, p.18.

Buyar, V.F., and Korneichuk, N.F., (1980) Transportnaya Problema i yo Reshenie (The Transport Problem and its Solution) LP, July, p.18.

Chemyanin, V.A., (1975) Bezchokernaya Trelyovka Lesa v Ustkomskom Lespromkhoze (Grapple Skidding in the *Ustkomski Lespromkhoz*) LELS, No.28, p.6.

Chernozhskaya, G.T., (1980) Lesosplav v Usloviakh Vodokhranilitz (Timber Floating Near Reservoirs) LP, March, p.9.

Chesnokova, K.N, (1977) Agregatni Mashin Derzhat Ekzamen (Aggregate Machines are Examined) LP, January, p.17.

Chuprov, N.P., (1975) Ekonomicheskaya Spelost i Vozrast Rubki (Economic Maturity and the Rotation Length) LKh, October, p.24.

Conquest, R., (1971) The Great Terror, Pelican Books, London.

Danshchikov, M.G., and Michaelov, V.A., (1977) Nuzhni li Promezhutochnie Skladi? (Is There a Need for Temporary Landings?) LP, April, p.8.

Demidov, A.S., (1978) Vozdeistvie Bolshegruznikh Lesovosnikh Avtopoezdov na Koleinoe Zhelezobetonnoe Pokritie (The Effect of Large Lorries on Concrete Road Coverings) VNIPIEILesProm Ek Info, No.11.

Deyev, B.A., (1978) Vnedrenie Agregatnikh Mashin na Lesozagotovitelnikh Predpriatiakh Srednovo Urala (The Introduction of Aggregate Machines in Logging Enterprises of the Urals) VNIPIEILesProm Ek Info, No.9.

Dmitrin, A.G., (1978) Podgotovka Rukovodyashchevo Zveno (Training of Administrative Personnel) LP, September, p.1.

Donski, I.P., and Savelov, V.V., (1973) Vodni Transport Lesa (Timber Floating) LPPH, Moscow.

Dyakonov, F., (1975) The Outlook for Development in the Soviet Far East, Current Digest of the Soviet Press, 30 April.

Eason, W., (1979) Demographic Divergencies at the Republic Level, in Regional Development in the USSR: Trends and Prospects Colloquium 25-27 April 1979, Oriental Research Partners, Newtonville, Mass., USA.

Eckholm, E., (1979) Planting for the Future: Forestry for Human Needs, Worldwatch Paper No.26, Worldwatch Institute, Washington DC.

Egorov, L., (1970) Traktor K-703 na Vivoskie Drevesini (The K-703 Extracting Timber) LP, November, p.10.

Elenski, S.I., and Korotishev, A.A.,(1976) Ekonomicheski Aspekti Proizvodstvenovo Travmatism (Economic Aspects of Industrial Injury) LP, January, p.26.

Ellman, M., (1971) Soviet Planning Today, Cambridge University Press.

Ellman, M., (1973) Planning Problems in the USSR, Cambridge University Press.

Emlyanov, V.M., (1972) Organizatsia Lesosechnikh Rabot na Baze Novoi Tekhniki (Organisation of Logging on the Basis of the New Technology) LELS, No 25, p.4.

Fedorenko, N.P., (1974) Optimal Functioning System for a Socialist Economy, Progress Publishers, Moscow.

Filippskikh, S.M.,(1978) Lesotekhnicheskaya Shkola i Proizvodstvo (Technical Forestry Schools and Production) LP, December, p.21.

Gatskevich, V.A., (1971) Spravochnik Mastera Lesozagotovok (A Logger's Handbook) LPPH, Moscow.

Gatskevich, V.A., (1977) S Chevo Nachinalas Mekhanizatsia (How Mechanisation Began) LP, October, p.10.

Ginsburg, V.L., (1978) Ritm Transportnovo Konveira (The Rhythm of Transportation) LP, September, p.10.

Ginzburg, Y.S., (1968) Into the Whirlwind, Penguin Books, London.

Girgidov, D. Ya, and Tostopyatenko, A.I., (1977) Sozdaniya Leso-syemennikh Plantatsii Khvoinikh Porod (Establishing Coniferous Seed Orchards) LenNIILKh, Leningrad.

Glotov, V.V., (1973) Optimalnoi Planirovanie v Lesnoi Promishlen-nosti (Optimal Planning in Logging) LPPH, Moscow.

Glotov, V.V., Urkin, R.V., and Ruzin, S.I., (1975) Ekonomika, Organizatsia i Planirovanie Lesozagotovitelnova Proizvodstva (The Economics, Organisation and Planning of Logging) LPPH, Moscow.

Glotov, V.V., (1977) Razmeshcheniye Lesopromishlennovo Proizvodstva (Location of the Forest Products' Industries) LPPH, Moscow.

Goffman, K.G., (1974) Uchot Faktor Vremeni pri Ekonomicheskoi Otsenke Lesnikh Zemel (The Time Factor in the Valuation of Forest Land) in Ekonomicheskaya Otsenka Lesnikh Zemel, Institut Ekonomiki Akademii Nauki Litovskoi SSR, Vilnius.

Gorbachevski, V.A., (1973) Avtomobilni Transport Lesa (Road Transport of Timber) LPPH, Moscow.

Gorbov, A.F., and Zabyalov, A. Ya.,(1977) Vakhtovi Metod Zagotovki Lesa na Predpriyatiakh Obyedenenie Archangelsklesprom (The *Vakhtovi* Method in *Archangelsklesprom*) VNIPIEILesProm Ek Info, No 18.

Goretskaya, E.N., (1978) Nadyoznost Gidrosystem Lesnikh Mashin (The Reliability of Logging Machines' Hydraulic Systems) LP, October, p.21.

Gregory, R.R., and Stuart, R.C., (1974) Soviet Economic Structure and Performance, Harper and Row, New York.

Grossman, P., (1978) Labour Supply: Constraints and Responses, in The Future of the Soviet Economy, ed. H. Hunter, Westview Press Boulder, Colorado.

Guleka, V., (1970) V Borbe za Snizenie Proizvodstvennovo Travmatisma (In the Battle for the Reduction of Industrial Injury) LP, July, p.12.

Hamilton, G.J., and Christie, J.M., (1971) Forest Management Tables, Forestry Commission Booklet No 34, HMSO, London.

Helgeson, A., (1979) Comment, in Regional Development in the USSR, Trends and Prospects Colloquium 25-27 April 1979, Oriental Research Partners, Newtonville, Mass., USA.

Ilin, V.A., (1973) Sukhoputni Transport Lesa (Land Transport of Timber) LPPH, Moscow.

Jasny, N., (1959) Labor and Output in Soviet Concentration Camps, Journal of Political Economy, vol 51, p.405.

Johnson, D.R., Grayson, A.J., and Bradley, A.T., (1967) Forest Planning, Faber and Faber, London.

Kanevski, M.V., and Voronitsin, K.I., (1974) Lesozagotovki v Gornikh Raionakh SSSR (Logging in Mountainous Areas of the USSR) LPPH, Moscow.

Kanevski, M.V., and Shaitanov, G.Ya.,(1975) Lesnoi Export SSSR (Timber Exports of the USSR) LPPH, Moscow.

Kanevski, M.V., (1979) Visokaya Otvestvennost Lesozagotovitelei (Great Responsibilities for Loggers) LP, February, p.1.

Kapitonov, I.T., (1952) Elektricheski Trelevochni Traktor (The Electric Skidder) LP, May, p.15.

Kapustin, V.A., (1970) O Konstruktsie VPM (On the Construction of Feller-bunchers) LP, June, p.10.

Khomenka, B.F., (1978) Sovershenstvovat Ekonomicheskovo Obrazovanie v Otrasli (Improvements of Economic Education in the Sector) LP, December, p.1.

Kirin, L.M., (1981) God Otkrivaushchi Pyatiletki (First Year of the Five-year Plan) LP, January, p.1.

Kirukhin, K.D., (1977) Perspektivi Osvoeniya Lesnikh Massivov v Zone Vliyaniya BAMa (Long-term Exploitation of the Timber Stocks in the Area of BAM) VNIPIEILesProm Ek Info, No 6, Moscow.

Knickerbocker, H.R., (1931) The Soviet Five-year Plan and its Effects on World Trade, The Bodley Head, London.

Kochegarov, V.G., Fedayev, A., and Lavrov, A., (1970) Tekhnologia i Mashini Lesosechnikh i Lesovostenovitelnikh Rabot (Technology and Machines in Logging and Forestry) LPPH, Moscow.

Konstantinov, A.R., and Struzer, L.R., (1974) Lesnie Polosi i Urozhai (Forest Belts and Harvests) Gidrometeoizdat, Leningrad.

Korablev, A.I., (1980) Drova, ne Otkhodi a Resursi (Firewood, not a Waste but a Resource) LP, August, p.1.

Korobov, G.B., and Ribakov, D.M., (1977) Mnogooperatsionnie Lesozagotovitelnie Agregati za Rubezom (Foreign Logging Machines) VNIPIEILesProm Ek Info, No 19, Moscow.

Koroleff, A., (1952) Logging Mechanisation in the USSR, Pulp and Paper Research Institute of Canada, Montreal.

Koroleff, A., (1959) Logging Mechanisation in the USSR II, Pulp and Paper Research Institute of Canada, Montreal.

Korotonoshko, H.I., and Shebalin, U.A., (1950) Parovoi Avtomobile i Parosilovaya Ustanovka dlya Lesnoi Promishlennosti (Steam Lorries and Steam Powered Equipment for Logging) LP, July, p.11.

Korshunov, V.V., (1977) Nasushchnie Zadachi Remontnikh Sluzhb (Urgent Tasks for Repairers) LP, June, p.8.

Kozhin, V.M., (1972) Ulushit Ekonomiku Proizvodstva Tekhnologicheskoi Shchepi (Improve the Economics of Chip Production) LP, February, p.6.

Kozhin, V.M., Vorakso, V.K., and Khorshavin, E.I., (1975) Tsenoobrazovania v Lesozagotovitelnoi Promishlennosti (Price Formation in Logging) LPPH, Moscow.

Kozhin, V.M., and Styazhkin, V.P., (1976) Sebestoimost, Tseni i Rentabelnost na Lesozagotovkakh (Costs, Prices and Profitability in Logging) LPPH, Moscow.

Krainov, V.P., Gilev, N.K., Pervykhin, V.D., Ivankovich, A.S., and Lashmanov, V.I., (1972) Effektivnost Mekhanizatsii Lesozagotovok (The Effectiveness of Logging Mechanisation) LPPH, Moscow.

Krestiashin, L.I., (1974) Lesovodsvennoekonomicheskoe Obosnovanie Optimalnovo Vozrasta Rubki Glavnovo Polzovania v Ekspluatatsionnikh Lesakh Severo-Zapada (The Forest Economic Basis of the Optimal Rotation Length for the Production Forests of the North-West) LenNIILKh, Leningrad.

Krupnik, L.S., (1975) Novi Oblik Lesnovo Poselka (A New View of a Logging Village) LP, September, p.20.

Kudrivtsev, Iu. A., (1975) Lesopilnaya Promishlennost SSSR: Sostoyaniye i Problemi yo Razvitia (The Sawmilling Industry of the USSR: Its Present State and Problems of its Development) Lesnoi Zhurnal, No 1, p.5, Archangelsk.

Kulakov, A.K., (1980) Ot Eksperimenta k Praktike (From Experiment to Practice) LP, June, p.16.

Kuleshov, M.V., (1981) Velikaya Sila Sorevnovaniya (The Great Power of Competition) LP, February, p.4.

Kulikov, V.N., Popov, F.P., Olenich, V.I., and Berezin, V.S., (1977) Problema Zapasnikh Chastie i yo Reshenie (The Problem of Spare Parts and its Solution) LP, June, p.9.

Kulikov, V.N., and Popov, F.P., (1979) Opornie Punkti i Nadyozhnost Oborudovanie (Information Gathering Points and the Reliability of Equipment) LP, October, p.8.

Laiko, M.V., (1950) Vishe Kachestvo Mashin dlya Lesozagotovok (Better Machines for Logging) LP, July, p.4.

Lazarev, A.S., (1979) Kachestvo Nasazhdenii i Rentabelnost (The Quality of Plantations and Profitability) LP, January, p.5.

Lebedev, K.I., and Pominova, G.I., (1974) Vodni Transport Lesa (Water Transport of Timber) LPPH, Moscow.

Liceman, V.A., (1978) Osvoenie Mashin LP-19 v Karabashkom Lespromkhoze (The Introduction of the LP-19 in the *Karabashki Lespromkhoz)* VNIPIEILesProm Ek Info, No 4, Moscow.

Lobovikov, T.S., and Petrov, A.P., (1976) Ekonomika Kompleksnovo Ispolzovania Drevesini (The Economics of the Complete Utilisation of Timber) LPPH, Moscow.

Lobovikov, T.S., (1977) Lesnie Predpriatia Budeshchevo (Forest Enterprises of the Future) Lesnoi Zhurnal, No 3, p.134, Archangelsk.

Loffler, H.D., (1978) Developments in Forestry Technology in Western Europe, World Wood, May, p.9.

Loiberg, M.Ya, (1978) Molodozh na Lesozagotovkakh (Youth in Logging) LP, January, p.5.

Lumenov, R.A., (1975) Mashini dlya Rubki Krupnomernikh Derevev (Machines for Felling Large Timber) LELS, No 1, p.4.

Maklukov, L.M., (1982) Okhrana Truda-Otvetstvennii Podkhod (Safety at Work - Crucial Approach) LP, February, 2.

Medvedev, N.A., (1976) Put k Visoki Effektivnosti (The Path to High Effectiveness) LP, March, p.1.

Medvedev, N.A., (1977) Reservi Effektivnosti (Reserves of Effectiveness) LP, May, p.1.

Minlesprom, (1971) O Sovershenstvovania Organizatsii Upravlenia Lesnoi i Derevoobrabativaushchi Promishlennosti (On the Improvement of Management in Logging and Woodworking) Minlesprom Prikaz No 220, Moscow.

Minlesprom, (1972) O Sozdani i Vnedreni System Mashin dlya Kompleksnoi Mekhanizatsii Proizvodstva v Lesnoi i Derevoobrabativaushchi Promishlennosti na Period do 1990 Goda (On the Creation and Introduction of Systems of Machines for the Complete Mechanisation of the Logging and Woodworking Industries for the Period until 1990) Minlesprom Prikaz No 337, Moscow.

Minonov, T., (1948) Ispitanie Elektropili EPKh-3 (Tests of the Electric Saws EPKh-3) LP, October, p.7.

Misbko, Z.Z., (1980) Zagotovka Lesa v Gornikh Usloviakh s Ispolzovaniem Agregatnikh Mashin (Logging in Mountainous Areas with Machines) VNIPIEILesProm Ek Info, No 2, Moscow.

Mitin, A.G., and Sarkisov, V.V., (1974) Ekonomicheskaya Effektivnost Vnedrenia Novoi Tekhniki i Kapitalnikh Vlozenii v Lesnoi i Derevoobrabativaushchi Promishlennosti (The Economic Effectiveness of the Introduction of New Technology in the Logging and Woodworking Industries) LPPH, Moscow.

Morner, B., (1976) Survey of Chain Saws for Logging, World Wood, July, p.28.

Moscvin, N.A., (1976) Oblegchaem Trud Zhenshchin (Let us Ease the Work of Women) LP, May, p.11.

Moshonkin, N.P., (1975) Ekonomiki i Intensifikatsia Lesozagotovitelnoi Promishlennosti (Economics and the Concentration of Logging) LPPH, Moscow.

Moshonkin, N.P., (1978) Puti Povishenie Effektivnosti Proizvodstva (Ways to Increase the Effectiveness of Production) LP, November, p.1.

Moskin, N.P., (1968) Ekonomicheskie Osnovi Perekhoda Predpriyati Lesnoi Promishlennosti na Novuyu Systemu Planirovani i Ekonomicheskovo Stimulirovania (The Economic Basis for the Conversion of Forest Enterprises to the New System of Planning and Economic Incentives) LPPH, Moscow.

Muzukin, V.S., (1951) Promishlennie Tok - na Lesozagotovki (The Use of Industrial Power Lines in Logging) LP, November, p.8.

Nemtsov, V.D., (1981) Novaya Pyatiletka: Gorizonti Mashinizatsii (The New Five-year Plan: Horizons of Machine Introduction) LP, January, p.6.

Nemtsov, V.P., Abol, P.I., and Vinogorov, G.K., (1979) Sovremeni Lesosechnie Mashini (Modern Logging Machines) LP, October, p.28.

Nikolaenko, V.T., Plotnikov, L.A., and Voronina, A.P., (1973) Lesa Pervoi Gruppi (Forests of the First Group) LPPH, Moscow.

Nikolauk, V.A., (1979) Izmenenie v Lesnom Fonde v Rezultate Khoziatvennoi Dyatelnosti (Changes in the Forest Growing Stock as a Result of Sylvicultural Measures) LKh, July, p.10.

Nikulichev, I.S., (1975) Bit - Problema Kompleksnaya (Living Conditions - a Complex Problem) LP, September, p.12.

North, R.N., and Solecki, J.J., (1977) The Soviet Forest Products Industry: Its Present and Potential Exports, Canadian Slavonic Papers, September, p.281.

Nove, A.,(1969) An Economic History of the USSR, Allen Lane, London.

Nove, A., (1980) The Soviet Economy: Problems and Prospects, New Left Review, January/February, p.10.

Novokharstkin, V.G., (1978) Obsluzdautsa Problemi Zhenskovo Truda (Looking at the Problem of Women Workers) LP, December, p.5.

Ogievski, V.V., (1974) Lesnie Kulturi i Melioratsia (Sylvicultural Operations) LPPH, Moscow.

Okhotski, V.A., (1976) Vnedrenie Dvukhsmennoi System Raboti (The

Introduction of Two-shift Work) LELS, No.34, p.5.

Omarov, A.M., (1974) Planirovanie na Promishlennikh Predpriyatiyakh (Planning in Production Enterprises) Misl, Moscow.

Orlov, S.F., and Kochegarov, V.G., (1973) Lesosechnie Raboti byez Ruchnovo Truda (Logging Without Manual Labour) LPPH, Moscow

Orlov, N.K., and Skomorokhova, T.M., (1981) Ot Novoi Tekhniki - Polnuyu Otdachu (From the New Technology, Full Output) LP, December, 4.

Ovchinikov, M.M., and Donski, I.P., (1978) Osnovie Problemi Vodnovo Transporta Lesa (Basic Problems of Timber Floating) LP, April, p.11.

Parfenov, G.M. (1948) Statsionarnie ili Perdvizenie Lokomobili (Portable or Stationary Steam Engines) LP, November, p.14.

Patyakin, V.I., (1973) Vodni Transport Lesa (Water Transport of Timber) LPPH, Moscow.

Pautkin, U.P., (1977) Suchkoreznie Mashini v Lespromkhozakh Prikamya (Delimbers in the *Lespromkhozi* of Prikamya) LP, January, p.23.

Perfilov, M.A., and Stupnev, G.K., (1968a) Valochno-trelevochnaya Mashina VTM-4 (The VTM-4 Feller-skidder) LP, July, p.10.

Perfilov, M.A., and Stupnev, G.K., (1968b) Rezultati Ekspluatatsionikh Ispitanii VTM-4 (Results of VTM-4 Trials) LP, August, p.8.

Perfilov, M.A., (1974a) Mnogooperatsionnie Lesosechnie Mashini (Multioperation Logging Machines) LPPH, Moscow.

Perfilov, M.A., (1974b) Opit Raboti VTM-4 (Experiences of the VTM-4 at Work) LP, October, p.11.

Petrov, A.P., (1971a) Labour and Wage Organisation in the Logging Industry of the USSR, Working Paper for the Institutt for Skogokonomi, Oslo, Norway.

Petrov, A.P., (1971b) Evaluation of Economic Efficiency in the Forest Industries of the USSR, Working Paper for the Institutt for Skogokonomi, Oslo, Norway.

Polekhin, B.P., and Belenov, I.A.,(1978) Splav Kruglikh Lesomaterialov Listvennikh Porod (Floating Roundwood of Broadleaved Species) VNIPIEILesProm Ek Info, No 5, Moscow.

Polekhin, B.P., and Dmitriev, I.N.,(1980) Zavisimost Plotnost Drevesini Sverzersrublennoi Listvennitsi ot Soderzania Zaboloni (The Relationship Between the Density of Freshly Felled Larch and the Sapwood Content) VNIPIEILesProm Ek Info, No 19, Moscow.

Popov, V.A., (1948) Itogi Blizaushie Perspektivi (Results and the Near Future) LP, May, p.1.

Popov, V.A., and Mikhaelov, V.A., (1975) Vnedrenie Traktora TB-1 v Obyedineniye Komilesprom (Introduction of the TB-1 in *Komilesprom*) LELS, No 28, p.3.

Reshetnikov, A.S., and Shugar, S.S., (1980) Vakhtovi Metod: Tekhnologia i Effektivnost (The *Vakhtovi* Method: Technology and Effectiveness) LP, January, p.8.

Reshetov, A.V., (1951) Potochno-kompleksnie Brigadi na Lesozagotovok v Sibirii (Complex Brigades in Logging in Siberia) LP, July, p.13.

Rodigin, A.A., Platonov, P.I., and Morozov, F.N., (1972) Organizizatsia i Planirovania Lesnoi Promishlennosti i Lesnovo Khoziastva (The Organisation and Planning of Forestry) LPPH, Moscow.

Rodigin, A.A., (1977a) Planirovania Proizvodstvennoi Programmi Lesekspluatatsia i Lesosplav (The Planning of Production of Logging and Timber Floating) Lesotekhnicheskaya Academia, Leningrad.

Rodigin, A.A., (1977b) Planirovania Truda, Fondov Zarabotoi Plati i Povishenie Proizvodstva (Planning Labour, Wages and the Increase of Production) Lesotekhnicheskaya Academia, Leningrad.

Rodigin, A.A., (1977c) Planirovania Sebestoimosti Produktsi i Rentablenosti (Planning Costs of Products and Profitability) Lesotekhnicheskaya Academia, Leningrad.

Rodigin, A.A., (1977d) Proizvodstvennofinansovi Plan po Lesnomu Khoziastvu (Sylvicultural Planning) Lesotekhnicheskaya Academia, Leningrad.

Rodigin, A.A., (1977e) Novaya Tekhniki i Tempi Rosta Proizvoditelnosti Truda (New Technologies and the Growth of Labour Productivity) LP, January, p.25.

Rogochev, F.I., and Shirshov, N.V., (1948) Vo Vladimirskikh Lesakh (In the Forests of the Vladimir Region) LP, September, p.2.

Roos, L.V., (1948) Trelevochni Traktor KT-12 (The KT-12 Skidder) LP, April, p.8.

Rumyantsev, G.T., and Konyeva, N.I.,(1972) Isledovaniye Ekonomicheskoi Effektivnosti Iskustvennovo i Yestyestvennovo Lesovozobnovleniye (Research into the Economics of Artificial and Natural Regeneration) in Isledovaniye po Lesnomu Khoziastvu, Lenizdat, Pskov.

Sakharov, V.V., and Sokikas, V.I., (1980) Shto Pokazal Ekspluatatsia Zarubeznoi Tekhniki? (What Does the Use of Foreign Machines Show?) LP, January, p.21.

Sanikov, U.G., Gorev, G.I., and Smolenkov, A.A., (1980) Kakovi Realnie Zapasi Pnevoi Drevesini? (What are the Real Reserves of Stumpwood?) LP, February, p.6.

Seperovich, I.P., (1976) Faktor Vremeni v Lesnom Khoziastve (The Time Factor in Forestry) LKh, March, p.30.

Seroken, N.A., and Filimonov, S.S., (1980) Okhrana Vodnikh Obyektov pri Lesosplav (Protection of Reservoirs when Floating Timber) LP, November, p.28.

Shahniashvili, E.E., (1951) O Komplektnom Primenenie Elektricheskoi

Energii na Lesozagotovkakh (On the Total Use of Electricity in Logging) LP, January, p.9.

Shalaev, S.A., (1982) Lesozagotovitelnaya Promishlennost v 1982 godu (The Logging Industry in 1982) LP, January, 1.

Shchepba, V.G., (1980) Vakhtovi Metod - Vyes God (The Vakhtovi Method - All the Year) LP, January, p.8.

Shcherbakov, V.A., (1978) Shvedskie Lesozagotovitelnie Mashini (Swedish Logging Machines) VNIPIEILesProm Ek Info, No 8, Moscow.

Shcherbakov, V.A., Tishkin, V.I., and Terkhanova, L.S., (1978) Opit Vnedrenie Vyesovo Metoda Uchota Lesomaterialov v Bisetskom Lespromkhoze (The Use of the Weight Method of Measurement of Timber in the *Bisetski Lespromkhoz*) VNIPIEILesProm Ek Info, No 2, Moscow.

Shimanuk, A., (1974) Dendrologia (Dendrology) LPPH, Moscow.

Shtrek, V.V., (1978) Ekspluatatsia Agregatnikh Mashin VM-4 i LP-18 v Obyedineniye Kranoyarsklesprom (The Use of the VM-4 and LP-18 in *Krasnoyarsklesprom*) VNIPIEILesProm Ek Info, No 4, Moscow.

Shtrek, V.V., Plemyaninkov, M.V., Sorokin, V.V., and Lozitski, B.S., (1980) Effektivnost Kompletka Mashin pri Viborochnikh Rubkakh (The Effectiveness of Machine Systems in Selective Cuttings) LP, June, p.19.

Shugar, S.S., (1978) Rabot Valochno-paketirushchi Mashini LP-19 i Pachkopodborshchikov LT-157 po Mnogosmennoi Rezim v Obyedineniye Tumenlesprom (The Work of the Feller-buncher LP-19 and the LT-157 Grapple Skidder in Multi-shift Working in *Tumenlesprom*) VNIPIEILesProm Ek Info, No 4, Moscow.

Sinyakevich, I.M., (1979) Usilit Stimuliruushchi Funktsii Tseni (Strengthen the Stimulating Function of Price) LP, February, p.19.

Siritov, I.I., Siritov, V.I., and Prokhorov, I.K., (1972) Lesoekpluatatsia (Logging) LPPH, Moscow.

Smirnov, S.D., (1977) Opit Sozdania Postoyannoi Lesosyemennoi Bazi (Experiences in the Creation of Permanent Seed Orchards) LPPH, Moscow.

Solomonov, V.D., (1981) Gorizonti Lesnoi Nauki (Horizons of Forestry Science) LP, February, p.22.

Solzhenitsyn, A., (1963) One Day in the Life of Ivan Denisovich, Penguin Books, London.

Stepanov, V.N., (1979) Zima Sposit Strovo (The Importance of Winter Operations) LP, October, p.4.

Stupnev, G.K., Khasdan, S.M., and Plakhov, V.N., (1976) Derevoobrabativaushchaya Promishlennost za Godi Devyatoi Pyatiletki (The Wood-working Industry in the Ninth Five-year Plan) LPPH, Moscow.

Stupnev, G.K., (1978a) Monogooperatsionnie Mashini; Poiski, Reshenie

(Multioperation Machines; Investigation, Decision) LP, February, p.5.

Stupnev, G.K., (1978b) Lespromkhoz Shagaet v Zavtra (Logging in the Future) LP, December, p.3.

Stupnev, G.K., (1979) Perspektivi Razvitia Kompleksnoi Mekhanizatsii i Avtomatizatsii Prosesov Proizvodstva v Lesnoi i Derevoobrabativaushchi Promishlennosti (The Long-Term Development of the Total Mechanisation and Automation of the Production Processes in Logging and Woodworking) VNIPIEILesProm Ek Info, No 4, Moscow.

Stupnev, G.K., (1981) Inzeneri Poisk - na Sluzbu Proizvodstvu (Engineering Research in the Service of Production) LP, February, p.20.

Sudev, N.G., Novikov, B.N., and Rozin, L.N., (1976) Lesokhoziastvennoi Spravochnik dlya Lesozagotovitelyei (Sylvicultural Handbook for Loggers) LPPH, Moscow.

Svirinuk, K., (1971) Opit Ekspluatatsii Suchkoreznikh Mashin (Results of the Use of Delimbers) LP, January, p.6.

Swianiewicz, S., (1965) Forced Labour in the Soviet Union, Oxford University Press, Oxford.

Tatarinov, V.P., (1978) Kak Razvivatsa Industrii Lesa (How the Logging Industry Develops) LP, November, p.1.

Tauber, B.A., (1979) Perspektivi Ispolzovaniya Vsyei Biomassa (Long-term Use of all the Biomass) LP, June, p.5.

Tausnev, A.B., (1979) Napravlenia Razvitia Mekhanizatsii Lesozagotovitelnova Proizvodstva v Obyedineniye Tumenlesproma (The Direction of the Logging Mechanisation in *Tumenlesprom*) VNIPIEILesProm Ek Info, No 4, Moscow.

Terentyev, A.I., and Soldatov, T.M., (1975) V Gostapriimnom Lesnom Krai (In the Gostepriimni Forest Region) LP, September, p.8.

Thornton, J., (1971) Differential Capital Charges and Resource Allocation in the Soviet Industry, Journal of Political Economy, vol. 19, p.545.

Timofeyev, N.V., (1975) Generalnaya Skhema Upravlenia Lesnoi Industri (General Plan of Management of the Forest Industries) Planovoe Khoziastvo, No 9, p.9.

Timofeyev, N.V., (1978) Rubezi 1978 (Goals for 1978) LP, January, p.1.

Traktinski, E.B., (1977) Pyatiletka i Problemi Podgotovka Rabochnikh Kadrov (The Five-year Plan and the Problem of Preparing the Work Force) LP, June, p.1.

Tseplayev, V.P., (1965) The Forests of the USSR, Israel Program for Scientific Translation, Jerusalem.

Tsibayev, N.M., (1980) Leto Otvetstvenni Etap Lesozagotovok (Summer - a Key Time for Loggers) LP, June, p.3.

Tsibulya, G.M., (1975) Uspyekh Rosdaetsa v Brigade (Success Appears in the Logging Brigade) LP, February, p.18.

Tsimek, A.A., (1975) Uchot Faktora Vremeni v Lesnom Khoziastve (The Factor of Time in Forestry) LKh, January, p.20.

Urkin, R., (1969) O Zatratakh na Ochistiki Lesosyek (On the Costs of Brash Clearance) LP, April, p.12.

Uvarov, N.V., (1950) Benzomotornie Pili - na Zagotovku Khlistov (Petrol Driven Chain Saws in Tree Length Logging) LP, June, p.20.

Varaksin, F.D., and Stupnev, G.K., (1974) Osnovniye Napravlenia Tekhnicheskovo Progressa Lesnoi i Derevoobrabativaushchi Promishlennosti (The Directions of Technical Progress in the Logging and Wood-working Industry) LPPH, Moscow.

Vasiliyev, L.V., and Gryazin, A.D., (1979) Pochemu Inzener Ukhodit iz Otrasli? (Who do Engineers Leave the Industry?) LP, February, p.7.

Vingorov, G.K., (1969) Ochistika Lesosyek v Sovremennon Aspekt (A Modern View of Brash Clearance) Trudi TsNIIME, No 96, Khimki Moscow.

Vingorov, G.K., (1972) Nekotorie Leso-ekspluatatsie Kharakteristiki Pochvenno-gruntovikh Uslovii i Relefa (Some Logging Characteristics of the Soil and Relief Aspects of Soviet Forests) Trudi TsNIIME, No 123, Khimki Moscow.

Vorobiev, G.I., Moiseyev, N.A., Lositski, K.B., Sinitsin, S.G., Tsekhnistrenko, A.F., Feofilov, V.A., Tsimek, A.A., Pobedinski, A.V., and Urkevich, M.S., (1979) Ekonomicheskaya Geografia Lesnikh Resursov SSSR (An Economic Geography of the Soviet Union's Forests) LPPH, Moscow.

Vorobiev, G.I., (1981) Novi Rubezi Lesnovo Khoziastva (Frontiers of Forestry) LKh, February, p.2.

Voronin, I.V., and Panitseva, L.I., (1976) K Voprosu Ucheta Faktora Vremeni v Lesnom Khoziastve (Towards the Question of Time in Forestry) LKh, March, p.26.

Voronitsin, K.I., (1957) Nauka na Sluzbe Mekhanizatsii Lesozagotovok (Science in the Service of Logging) LP, October, p.9.

Voronitsin, K.I., Lakhno, V.P., Lepentsov, P.A., and Perfilov, M.A., (1972) Osnovnie Parametri i Tipaz Bazovikh Traktorov dlya lesozagotovitelnoi Promishlennosti na Perspektivi (Basic Characteristics of the Tractors for the Logging Industry for the Long-term Future) Trudi TsNIIME, No 123, Khimki Moscow.

Voronitsin, K.I., (1975) Zenshchina v Lesa; Kak Oblegchit yo Trud (Women in Forestry; How Can Their Work be Eased?) LP, December, p.4.

Vozni, V.P., (1972) Novaya Tekhnika dlya Lesozagotovok Sibiri (New Logging Technology for Siberia) LP, March, p.21.

World Bank, (1978) Forestry Sector Policy Paper, World Bank, Washington.

Yagudin, U.N., and Gorbachev, N.N., (1976) Vakhtovi Metod Lesozagotovok (The *Vahktovi* Method of Logging) LPPH, Moscow.

Yanushenko, A.P., (1968) Faktor Vremeni i Produktsia v Lesokhoziastvennoi Proizvodstva (The Time Factor in Forestry) in Voprosi Ekonomiki Lesnovo Khoziastva, Tsentralnochernozemnoe Izdatelstvo, Voronezh.

Yanushenko, A.P., (1977) Ekonomika Lesnova Khoziastva (The Economics of Forestry) Visheishaya Shkola, Minsk

Zaitsev, B., (1978) Determining the Effectiveness of Technical Innovations, Problems of Economics, vol. 21, No 2, p.23.

Zalegaller, B.G., and Lastochkin, P.V., (1973) Mekhanizatsia i Avtomatizatsia Rabot na Lesnikh Skladakh (Mechanisation and Automation of Work at Forest Landings) LPPH, Moscow.

Zima, I.M., and Matugin, T.T., (1976) Mekhanizatsia Lesokhoziastvennikh Rabot (Mechanisation of Sylvicultural Operations) LPPH, Moscow.

Zheltov, E.M.,(1977) Meropriatia po Perevodu Zhenshchin na Bolyee Lyodkie Vidi Rabot (Improvements in the Transfer of Women to Lighter Work) VNIPIEILesProm Ek Info, No 8, Moscow

Zverev, V.F., (1978) Tekhnicheskoe Pereoruzenie Otrasli (Technical Up-grading of Logging) LP, February, p.1.

Index

For Product Safety Concerns and Information please contact our EU representative GPSR@taylorandfrancis.com Taylor & Francis Verlag GmbH, Kaufingerstraße 24, 80331 München, Germany

Batch number: 10725905

Printed by Printforce, the Netherlands